City Behind A Fence

THE UNIVERSITY
OF TENNESSEE PRESS
KNOXVILLE

CITY BEHIND A FENCE

Oak Ridge, Tennessee, 1942–1946

Charles W. Johnson
and
Charles O. Jackson

Cloth: 1st printing, 1981; 2nd printing, 1986;
         3rd printing, 1993.
Paper: 1st printing, 1981; 2nd printing, 1982;
         3rd printing, 1986; 4th printing, 1987;
         5th printing, 1990; 6th printing, 1992;
         7th printing, 1999; 8th printing, 2002.

The paper used in this book meets the minimum requirements of
ANSI/NISO Z39.48-1992 (R 1997) (Permanence of Paper). The
binding materials have been chosen for strength and durability.

Library of Congress Cataloging in Publication Data

Johnson, Charles W.   1934–
      City behind a fence.
      Bibliography: p.
      Includes index.
      1. Oak Ridge, Tenn.—History. I. Jackson, Charles O., 1935–
joint author. II. Title.
F444.03J63   976.8'73   80-15897
ISBN 0–87049–303–5 (cl.: alk. paper)
ISBN 0–87049–309–4 (pbk.: alk paper)

Dedication

*This book is dedicated with love and thanks to my
mother, Mrs. Josephine Johnson, and to the memory of
my father, Luther Eric Johnson.*

CHARLES W. JOHNSON

*To Lois Stevens because she is my mother and has
worked hard at it. To Phyllis Cole because she is a lovely
lady who, I am sure, has always been very special
without ever working at it.*

CHARLES O. JACKSON

Contents

Illustrations

Abbreviations

AIT	American Industrial Transit
CEW	Clinton Engineer Works
CFAC	Central Facilities Advisory Council
DSM	Development of Substitute Materials—early name for MED project
FB&D	Ford, Bacon and Davis Construction Company
FPHA	Federal Public Housing Authority
FSA	Federal Security Administration
K-25	Gaseous Diffusion Separation facility
MED	Manhattan Engineer District
NDRC	National Defense Research Council
OPA	Office of Price Administration
ORRWA	Oak Ridge Recreation and Welfare Association
OSRD	Office of Scientific Research and Development
RA	Roane-Anderson Company
R&WA	Recreation and Welfare Association
S-1	Early name for MED project
S-50	Liquid Thermal Diffusion Separation facility
S&W	Stone and Webster Engineering Corporation
SED	Special Engineering Detachment

SOM	Skidmore, Owings and Merrill—Boston architectural firm
TDU	Temporary Dwelling Unit
TVA	Tennessee Valley Authority
USED	United States Engineer Department
X-10	Atomic Pile and Laboratory facility—later Oak Ridge National Laboratory
Y-12	Electromagnetic Separation facility

Acknowledgments

Bringing this volume to publication has involved not merely the best efforts of the authors but also substantial cooperation and assistance from a number of individuals and organizations. We should like to include a word of thanks for that help. Needless to say, anything other than a superficial study required full access to Oak Ridge community records. These materials were principally available at three locations: the Federal Records Center, East Point, Georgia; the National Archives, Washington, D.C.; and the Department of Energy, Oak Ridge, Tennessee. Personnel at all three points were extremely helpful in directing the authors to appropriate sources.

Particular appreciation is due to four employees of the Department of Energy. The authors greatly benefited early in the present project from the advice and suggestions of agency historian, Dr. Richard Hewlett. Mr. Fred Peitzsch, community specialist of the agency's Oak Ridge office traveled the extra mile to facilitate and encourage our research effort. Mr. Floyd Beets, records officer, and Mr. Frank Hoffman, chief of photographic services both also employed by the Department of Energy, Oak Ridge, were helpful in identifying pertinent material of several sorts. The cooperation and assistance of the agency as well as these individuals does not imply, of course, any endorsement of this volume.

A number of colleagues at the University of Tennessee read the work at various stages in the manuscript's development and provided us with valuable suggestions for improvement. While all were most helpful, the following deserve special mention: Professors June Adamson, LeRoy Graf, Ralph Haskins, and Samuel Wallace. Three other individuals at the University sity warrant our gratitude for their assistance in physical preparation of the manuscript, repeated proof reading, and undertaking tasks which freed the authors to attend to the process of revision and rewriting. They are Mrs. Phyllis Cole, Ms. Susan Caolo, and Ms. Karen Thornton. Aside from this individual help, we should also like to acknowledge official aid from the University in the form of two summer-salary research grants to Dr. Johnson, and one research-assistance grant to Dr. Jackson.

We should like to thank Mr. Richard D. Smyser, editor of the *Oak Ridger,* for the time he willingly provided to our research effort. We must also extend thanks to a group of that city's residents, whose names we have chosen to keep in confidence, but without whose cooperation this volume would be much the poorer, if it could have been completed at all. These were the wartime residents of the reservation who agreed to oral interviews about that experience. They allowed us, of course, more than a conversation. We were provided entry into a very personal memory world of a highly significant part of their lives. We came away from these interviews with a genuine sense of privilege and warm appreciation for these gifts.

Part of the material in the Introduction and Chapters I, IV, and V of this work appeared earlier in our article entitled "The Urbane Frontier: The Army and the Community of Oak Ridge, Tennessee: 1942-1947," *Military Affairs,* February 1977, pp. 8–14, copyright 1977 by the American Military Institute. This material is reprinted here with permission, and no additional copies may be made without the express permission of the authors and the editor of *Military Affairs*.

Our deepest obligation, in every possible way, is to two very special ladies, Joan Milner and Emma Jackson.

Introduction

This is a history of Oak Ridge, Tennessee, during the years that top secret community played such a crucial role for what became America's entry into the Atomic Age. The deadly introduction to that era came in the massive destruction of the Japanese cities of Hiroshima and Nagasaki. The story of the development of the world's first nuclear bomb originated under circumstances much less well known than its fiery conclusion in 1945. The story began, insofar as any complex series of events can "begin," with two occurrences which pre-dated formal United States involvement in World War II.

The first episode took place in Germany in 1939 when scientists at Berlin's Kaiser Wilhelm Institute succeeded in splitting atoms of uranium by bombarding them with neutrons. Nuclear physicists had long understood the structure of the atom and had believed that atomic fission was theoretically possible. Fission was now accomplished, and it rapidly became clear that this new energy source could be used to produce a bomb more powerful and terrifying in magnitude than the world had ever dreamed. A number of physicists in America and England, many of whom had been forced into exile by Nazi and Fascist oppression, were soon deeply concerned that such a weapon would be produced by Hitler's regime and subsequently would endanger the freedom literally of the entire world. Led by Professor Enrico Fermi of Columbia University and Professor Albert Einstein of the Institute for Advanced Study at Princeton, these scientists urged President Franklin D. Roosevelt to initiate an American nuclear research program. Shortly thereafter a small program was begun.

The second occurrence in the development of the atomic bomb in the United States took place in 1940. By this point it was all too clear that in many areas, American science lagged behind that of Germany. With the prospects of U.S. involvement in World War II increasing, Roosevelt sought to mobilize American scientists toward a closing of that gap through creation of the National Defense Research Council. It included representatives from the federal government, universities, and private industry. In a reorganization of the national defense program in June 1941, Roosevelt subsequently created the Office of Scientific Research and Development. One primary concern of these bodies was the feasibility of developing a nuclear weapon.

The problems involved in producing an atomic bomb were monumental. One was to find a fissionable element available in adequate quantity. The uranium isotope U-235 was a possibility. But uranium was a rare metal, not easily extracted from its ore, and in its natural forms, there was only one-part Uranium-235 to 140 parts of the predominant U-238. The task of separating the vital U-235 isotope and accumulating it in amounts necessary for a bomb was staggering. Another possibility was to convert the uranium element U-238 into a new element, plutonium, which was as fissionable as U-235. That this could be accomplished was demonstrated in December 1942, when scientists at Chicago actually obtained a controlled uranium chain reaction. Still, the difficulties of applying this new knowledge to large-scale production of plutorium were enormous.

By summer of 1942, however, American nuclear

research had advanced sufficiently to insure that a full-scale bomb development program was feasible and advisable. This judgment, along with the recommendation for initiation of such a project, was transmitted to FDR on June 13, 1942, by Dr. James B. Conant, chairman of the National Defense Research Council, and Dr. Vannevar Bush, director of the Office of Scientific Research and Development. The report was approved by the President four days later. Shortly thereafter the Manhattan Engineer District (MED) was organized within the U.S. Army Corps of Engineers. Colonel (later Major General) Leslie R. Groves was chosen to command the secret project. The mission of the Manhattan program, for security purposes briefly known as Development of Substitute Materials, was the production of an atomic bomb within a three-year period.

The Conant-Bush report also urged that a search for an appropriate production site or sites be undertaken immediately. What became the MED reservation in Tennessee, known as Site X, was one of three major locations in the effort to move from nuclear theory to an operative atomic weapon. A second was at Hanford, Washington, on the Columbia River, a location known by MED personnel as Site W. This site, identified in December 1942, was in land area by far the largest of the three. It was to be the major production source for plutonium, the material which would comprise the so-called "Fat Man" bomb dropped on the Japanese city of Nagasaki in 1945. The third and most secret of the installations, code named Site Y, was on a lonely mesa at Los Alamos, outside Sante Fe, New Mexico. This location overlooking the upper Rio Grande Valley was the smallest of the three in size as well as population. It contained, however, the best equipped physics research laboratory in the world. Here, under the direction of Dr. J. Robert Oppenheimer of the University of California, actual design and construction of bombs took place with material obtained from the other two MED locations.

The Tennessee project was, in terms of the variety of operations carried out there, the most complex. Its primary mission was production of U-235, which was obtained through two separation methods: a gaseous diffusion and an electromagnetic process. A third though lesser effort for attaining U-235 through a thermal diffusion method was authorized at the project in 1944. Although plant facilities were completed for this process, it was later discontinued. A fourth method of separation, a centrifuge process, was originally considered but later rejected on the grounds of cost, complexity, and lack of promise. Ultimately, material produced in Tennessee became the base for the world's first use of atomic energy as a military weapon—the August 6, 1945, strike on the city of Hiroshima. The Tennessee location also included an atomic pile or graphite reactor that produced small quantities of plutonium. Finally, Oak Ridge was in population the largest of the three support communities and the most intricate in community organization. It was the first to be established and to some extent would serve as a model for operation of the Hanford site. In a brief three-year period it would also become the fifth largest city in Tennessee.

Much has been written about actual production of the atomic bomb, the technical activities at each of the three major development locations, as well as the military use of nuclear energy against Japan. Virtually no effort has been made, on the other hand, to examine the people and their lives at any of the three support communities. It is a curious oversight if only because successful operation of the communities was so crucial to the successful conclusion of the Manhattan project's atomic mission. Yet beyond this matter, these "secret cities," each built from the ground up, represent fascinating episodes in American social history. To chronicle the nature of human life in any of them is to report on a distinctly unique setting and chapter in the national past. All three communities well deserve scholarly attention. It is hoped that this book provides the needed statement on the Oak Ridge experience. That work remains to be done on Hanford and Los Alamos.

Given the carefully designed nature of the town, the authors of the present volume have been tempted to seek analogy between wartime Oak Ridge and a variety of past planned communities, utopian or otherwise. In these terms Oak Ridge could, for example, be compared with the New Deal greenbelt communities or even the Tennessee city of Kingsport, which was planned and developed one hundred miles to the northeast of Oak Ridge some twenty-five years earlier by private enterprise. But such analogy in the end was not appropriate. The military never believed they were building an ideal community, nor for that matter even a permanent one. The entire reservation presumably would exist only for the duration of the war. Moreover, the town was always of secondary interest to the Army. If there were niceties of life beyond the absolute necessities built into the community, the justification was primarily that such things would keep residents happy, morale high, and job turnover low. These amenities thus came directly from the paramount goal of completing the military mission successfully.

The authors were also tempted to make the more obvious comparison between the reservation townsite and a military post. As will be noted, however, the scope and complexity of the civilian role in the community operation were far too extensive to make such analogy a meaningful one. Indeed, even the Army rejected the comparison. A third contemplated possibility, and perhaps a more satisfying one, was analogy to the traditional company town with MED authorities filling the role of company management. Not unlike these towns, the reservation community was strictly planned and totally owned by the government. It was also occupation related. In keeping with the company town comparison, loss of employment meant loss of all right to remain in the town.

Yet considered further, the analogy proved clearly a weak one. The political, social, and economic life of the town was always much richer and more complex than in the typical company town. There was too much in the community which remained unplanned, and the city lacked the stability of most company towns. Finally, and quite dissimilar from the company town model, a consistent goal held by the military was to reduce as much as possible its direct participation, or obvious participation, in townsite operation and subsequently to enlarge as much as possible the civilian role. As will be shown, however, production and security needs put very severe limits on what the Army deemed possible in this realm.

Figuratively, if not literally, the best analogy which the authors concluded might be applied to Oak Ridge was that of the frontier community. Certainly the townsite looked like a frontier setting throughout the war years. The gravel roads, wooden sidewalks, constant construction, dust, and eternal mud would all give that impression. MED officials often used this pioneer analogy themselves, apparently convinced that (whether accurate or not) it was an effective device for urging residents to live more cheerfully under difficult conditions. Moreover, most residents would come to believe it. "Somehow there is a touch of the Klondike about the place and the people," as one resident put it. Characteristic of most frontier settings, the population of the town was noticeably young. Equally characteristic, all normal social, economic, and municipal institutions on the area would have to be created from the ground up. On the other hand, the obvious weakness of the analogy was that no frontier community ever approached the complexity and sophistication of the townsite. Nor have most towns ever experienced anything like the all pervasive presence of the military that characterized reservation life.

In final verdict the authors concluded that no analogy was in fact adequate, nor did there exist any meaningful historical precedent for the Oak Ridge experience. Oddly enough it was this very suspicion, first voiced by the authors in their shared office at the University of Tennessee, Knoxville, during an afternoon "bull session," which set us to the task and the challenge of exploring that community's early history. "What must have been the nature of this 'secret city' and what did it mean to live there in the war period," we began our discussion. That day now seems long ago, but with relative certainty we believe we have arrived at an answer. The story involved is an extraordinary and totally intriguing one. If the narrative which follows does not convey this, the authors freely admit with even more certainty that they have failed somehow to do justice to it.

City Behind A Fence

1. The Government Has Built a Village

The decision to locate a nuclear development site in East Tennessee was made personally by Leslie Groves on September 19, 1942, only two days after he had been named commander of the Manhattan project. In most ways Groves was an ideal choice. An outstanding West Point graduate in engineering, he was a career soldier cut from the same mold as Douglas MacArthur and George Patton. His weaknesses were few. A Federal Bureau of Investigation check turned up only his passion for chocolate candy (he would later store his supply in the same safe where the project's most important secrets were kept) and his concern for middle-age spread. It was true that Groves was hardly the most tactful officer to whom the government might have turned. His personality could be abrasive and his decisions often appeared arbitrary. Conversations or arguments which he regarded as without value were promptly concluded with a crisp "Enough!" He generated awe, if not outright fear in subordinates, and far too often ill temper among his equals.[1]

Yet in terms of capacity and ability to get things done few could surpass Leslie Groves. Forty-eight years old when he assumed command of the Manhattan project, he was tough, tireless, and relentless. He worked fifteen hours a day and expected similar dedication from those around him. He possessed a vocabu-

3

lary capable of blistering a construction worker and an ego which constantly assured him that his every decision was the correct one. As one of his few friends observed, "Groves not only behaves as if he can walk on water, but as if he actually invented the substance." He did not suffer fools easily, and by his disciplined pragmatic standards there seemed to be many, especially among the scientists of the Manhattan Engineer District (MED). Forever tinkering, insisting on yet one more test before acting, refusing to move on before one more modification was made, he disliked them collectively and in substantial numbers individually. Many of them felt equally negative toward Groves with his constant bullying and demands. Yet if hardly beloved, he was effective. The project did go forward on time and virtually every aspect of it, from the site location in Tennessee to the final details of the bombing run at Hiroshima, bore the general's personal imprint.[2]

With respect to the decision to establish a MED reservation in Tennessee, Groves had acted quickly but by no means precipitously. The search for an appropriate location had gone on much of the summer of 1942 and in very careful fashion. At this stage in planning, the idea was to place all four of the envisioned production plants (pile, electromagnetic, gaseous diffusion, and centrifuge) at one site. Presumably, this would achieve the most rapid and economical construction as well as afford maximum coordination and control of the development effort. It was later decided to discard one of the production techniques, the centrifuge process, and to locate elsewhere the giant atomic pile from which plutonium was produced. But with the single site as a goal, the requirements of all four plants influenced thinking throughout the search for a suitable location.

Prior to the final selection of East Tennessee, several other areas were given serious consideration. Two sites in the vicinity of Chicago were examined but ultimately determined to be too small for the location of

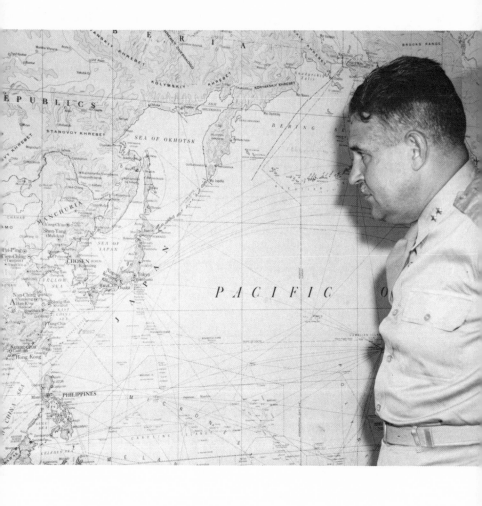

General Leslie R. Groves was the first commander of the
Manhattan project.

all four plants. The searchers also inspected a site near Shasta Dam in California but rejected it because it was too near the coast and therefore potentially vulnerable to enemy air attack. Finally, several areas in the state of Washington were studied. Each had the merit of availability of large amounts of electrical power from Bonneville and Grand Coulee dams. The areas were finally rejected because actually to obtain the needed power would require elaborate and time-consuming installation of long transmission lines, and, not unlike the California site, these areas might also be open to air attack.

From the standpoint of production activities the location identified in East Tennessee, however, was clearly superior to the others, in fact almost ideal. The isolated nature of this area would help minimize public awareness of the project's existence, and the site was far enough inland to be relatively safe from enemy attack. Since the potential dangers from the plants were not totally clear, the MED installation preferably should be located away from populous centers, and such was the case in Tennessee. Indeed, the actual terrain at the site was excellent in that operational dangers could be further minimized by placing plants so that they were separated by natural barriers.

The centrifuge, the electromagnetic, and the gaseous diffusion processes required large amounts of electricity, and Tennessee Valley Authority hydroelectric plants at Norris Dam, just north of the area, and Watts Bar to the south could provide much of that need. The remaining requirement was met ultimately through a power plant constructed by the Corps of Engineers. Necessary water of good quality was available at the site, a railroad link was nearby, and the area was accessible by motor transportation. Land was plentiful and cheap. It was estimated that the 52,000 acres (later enlarged to 59,000) necessary for the project could be obtained for an average of fifty to sixty dollars per acre. Displacement of those living in the region was

Map 1. Oak Ridge in Geographical Focus

considered to be a limited problem because the move would involve only about 1,000 families. Finally, the South in general was viewed as a favorable location, for it was a large reservoir of untapped non-farm labor that could be recruited to the project. Knoxville's 111,000 people would be a beginning point for local labor supply.[3]

Land acquisition began in the fall of 1942. Ultimately the total reservation would be approximately seventeen miles long east to west and would average seven miles in width. Tennessee's Clinch River meandered in such a way as to provide most of the eastern and western boundaries and the entire southern boundary of the project. In the beginning the area was so remote that only one paved road passed through it and MED employees arriving on the scene had to obtain water by truck from the nearby town of Clinton. The first telephone exchange for the ultra-secret project was established on the second floor above the Blue Moon Cafe. Originally disguised under the name Kingston Demolition Range, the reservation was officially designated as the Clinton Engineer Works early in 1943.

The gaseous diffusion plant, begun in the fall of 1943, was located on the reservation's northwestern perimeter. The principle of this process involved combining uranium with fluorine in a gaseous form and forcing the gas through a series of carefully designed perforated barriers that would cause the lighter U-235 to separate from the heavier U-238 isotopes. Ultimately the plant would be operated by Union Carbide and would carry the code name K-25. The letter K was derived from the first letter in the name of the plant's designer, Kellex Corporation. The number 25 was a common designation on the project for U-235 and was added arbitrarily.[4]

To the southeast of K-25 in a parallel valley would be the atomic pile, the pilot plant on which the huge plutonium production unit at Hanford, Washington, would be modeled. It was begun in February 1943. The

Union Carbide operates the massive K-25 gaseous diffusion facility.

pile was essentially a solid mass of graphite pierced by tubes running back and forth through it. When Uranium 238 was placed in the tubes in certain geometric designs, nuclear fission occurred. The result was transmutation of the uranium into small portions of plutonium. This plant was designated X-10, the symbol being essentially an arbitrary one. From its completion until July 1, 1945, X-10 was operated by the Metallurgical Laboratory of the University of Chicago. It was then taken over by the Monsanto Chemical Company. Further to the east and nearest to the townsite was the giant electromagnetic plant. Construction there also began in February 1943. By this process uranium was whirled through a magnetic field wherein the lighter U-235 atoms would spin out on a slightly different path than that of the heavier U-238 atoms. The two elements could then be separately gathered. This location would be identified as Y-12, the symbol having no particular significance, and it would be operated by the Tennessee Eastman Corporation (TEC).[5]

Most of the reservation was enclosed by a barbed-wire fence. In an unusual act of economy, the military salvaged much of that barbed wire from existing fences that had been installed by pre-MED residents. Entrance to the reservation was by one of seven gates. Three of these—White Wing, Gallaher, and Blair (Poplar Creek)—provided direct access to the "prohibited" or plant areas. The remaining four—Oliver Springs, Elza, Edgemoor, and Solway—allowed access to the townsite and the administrative area. The gate most immediate to the townsite was Elza, located at the eastern end of the project on what was originally Tennessee Route 61. Admission from the administrative and townsite areas to the "prohibited" areas would be controlled by four additional checking points.[6]

The townsite itself was placed in the northeast corner of the project on the slope of Black Oak Ridge some ten miles directly up the valley from the gas-

The graphite reactor is in the large central building shown
in this aerial photo of X-10 in March 1944.

eous diffusion plant. At its peak size the community extended down the north side of Route 61 from Elza Gate for about five miles and averaged in width perhaps a mile. On the south side of Route 61 approximately two miles from the Elza Gate were the administrative buildings. Slightly over one mile farther west was to have been the limits of the permanent residential community before the project employment demands destroyed that original design. The area just adjacent to that point both north and south of the highway would be given over to housing for the more transient construction workers. The nature of that housing would be largely trailers and racially segregated hutments, 16′ x 16′ single-story buildings that could accommodate four to six individuals.

In the early planning, responsibility for development of the townsite was originally given to the Stone and Webster Engineering Corporation (S&W), one of the prime contractors in the plant areas. District Engineer Colonel James C. Marshall and his staff, however, had doubts about Stone and Webster's overall town plan and especially about the plans for homes drawn up by the firm. Evaluation of these plans in January 1943 by Wilbur Kelly, principal engineer for the Corps, confirmed those doubts. Stone and Webster, he noted, had shown no originality or innovation in its basic town design, and Kelly believed the firm had seriously underestimated actual costs of constructing the houses, in some cases by over $1,000 a unit. There were mandated federal maximum amounts which might be spent per housing unit, and, if Kelly was correct, these could be met only at the sacrifice of space and "the few niceties" originally included. Although plant construction and operation were the primary purposes of the project, housing, Kelly argued, was not a sideline and should not be treated as such. House plans equal to or superior in quality to those of Stone and Webster could have been obtained, in Kelly's some-

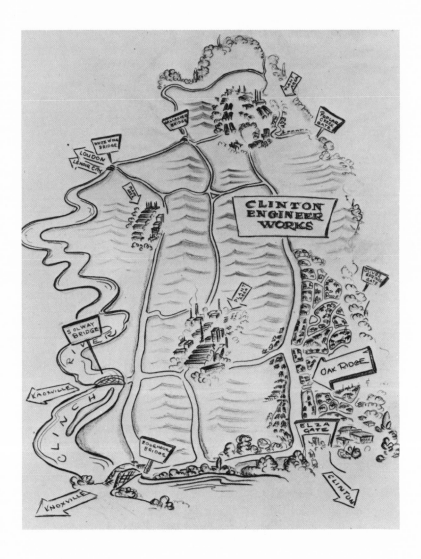

This wartime sketch locates most of the major facilities (X-10 and Y-12 are in the center).

what sardonic opinion, from any lumber yard at the cost of printing.[7]

In specific regard to housing design, MED officials had two alternatives: either force a reorganization in s&w's housing section to a more acceptable plan or reject the s&w plan completely and seek the assistance of some architectural firm experienced in low-cost housing work. Kelly's recommendation, not surprisingly, was to look elsewhere.[8] This move would allow Stone and Webster to devote full attention and efforts to plant design. Subsequently MED officials turned to the John B. Pierce Foundation of New York and the Boston architectural firm of Skidmore, Owings and Merrill. The Pierce Foundation had developed durable, inexpensive, prefabricated housing that seemed well suited for the reservation. This firm, primarily a research organization, acted as a consultant to Skidmore, Owings and Merrill who laid out the actual townsite design and supervised construction. Initial contact between the military and Skidmore, Owings and Merrill well epitomized the kinds of problems that civilian firms met in dealing with the Corps during the war.

The company was cooperative and showed interest in the opportunity to develop a town. "What size," queried the company? The military became uneasy. "Plan for 13,000," said the Army, "and if it needs to be bigger you will be informed." "Where is the town to be located?" "That information is classified," replied the Army. "Just prepare the plans and if they look promising, details will be forthcoming." Of course the firm could not proceed at all without information on terrain and topography. The Corps recognized this and so, grudgingly, provided a few aerial photographs as well as limited topographical maps, all with titles and names cut out. John Merrill returned with a preliminary design. The Army at least seemed pleased. "How soon could a site team visit the area," asked Merrill? "Immediately," said the Army and then promptly refused to divulge the team's destination. At length a compromise was reached. Six company architects, including Merrill, agreed to go to New York's Pennsylvania Station on a given day at a precise time. There they would be met by an MED representative who would direct them aboard a train and provide tickets in sealed envelopes. Only once they were on the train could they open the envelopes and learn where they were going.[9]

The actual decision on the Pierce Foundation-Skidmore, Owings and Merrill plan for the design of the town versus the original Stone and Webster plan was made at a conference on February 16, 1943. Among those present was Leon H. Zach, a Harvard-trained landscape architect who had worked very successfully on planning divisional cantonment lay outs for the Army and on the site plans for Camp Leonard Wood, Missouri. He had been temporarily assigned to Colonel Marshall by Colonel J.H. Stratton, chief of the Engineering Branch, Construction Division. Zach's criticisms and recommendations that afternoon would shape much of Oak Ridge's future.

The old Stone and Webster plan, he said, was composed of different and unrelated segments, with poor traffic circulation and little consideration to auto parking. Some roads were laid out at right angles in relation to the topography, making for excessive grades and others were laid out in such a way as to allow placing of houses only on one side. The plan paid small consideration to drainage. The Civilian War Housing Authority recommended a distance of at least 250 feet from the centerline of streets on one side of residential blocks to the centerline of streets on the opposite side. Many of the blocks in the S&W design did not meet this minimum. Finally, the conventional block treatment in the eastern section of town was not consistent with the more informal arrangement in the rest of the area.[10]

Although some modifications would have to be made, the Pierce-Merrill plans seemed well thought out. Some through streets and loops would have to be realigned and some blocks enlarged, especially where houses of a larger type had been turned at an angle to the street, but, in general, these plans showed much more thought and originality. The decision was made. Subsequently Zach reviewed his observation on the plan with John Merrill. The latter, who soon moved to Oak Ridge, would take a deep and personal interest in the city and its development.

The planners and District Engineer Marshall agreed on the following division of labor: Stone and Webster (as prime construction contractor) would lay in the utilities, but the Pierce-Merrill plans with some modification would form the basis for the town. Actual road construction would begin by the end of February 1943. Pierce would proceed as quickly as possible to produce a "1,000 house lay-out," including shopping and recreational facilities, using survey parties and draftsmen already in the field, and would submit a construction schedule on the first thousand houses that provided for a completion date of July 1.

Under the plan as agreed upon in February 1943, the residential area's arterial roads would have sidewalks on one side only. Each house would have a parking space, but there would be no garages. Block size would conform to a minimum of 250 feet between street centerlines with individual lot frontages of forty feet plus the house width. Twelve feet was the minimum distance from the house to the street. The houses were to face the inside of the block and be located so that the service yards for each two houses would be adjacent. Curb and gutter construction would be used for drainage rather than ditches because of the estimated higher maintenance cost of keeping ditches open in the steep grades found on the side of Black Oak Ridge.[11]

While Skidmore, Owings and Merrill was given substantial latitude in townsite design and construction, the Manhattan District planners had distinct ideas about many aspects of the community. The first, and one that was dealt an early crushing blow by employment demands in the plants, was that all operating personnel would be housed on the site.[12] Another was that resident comfort and attractiveness of surroundings should be matters of primary concern. Indeed, one reason for the interest in the Pierce Foundation's housing designs was that they could be constructed for considerably less than the $7,500 statutory maximum established for homes on military reservations. The amount saved in construction could be used to increase "livability" inside the house. Clearly, what the Corps did not want was the military camp model, which they themselves could have developed at much less cost. Top Corps officials actually visited locations where civilians were housed in government-built homes seeking to take their best characteristics and apply them at Clinton Engineer Works. It was after one such trip to Ocala, Florida, that Colonel Marshall determined to assure that extensive attention would be given to landscape design, including preservation of trees wherever possible.[13] That comfort of residents was a major concern is also borne out by Marshall's decision in the original design of single-family homes to equip each with a fireplace and porch, neither a necessity in the moderate climate of Tennessee.[14]

Clearing of land on the reservation got underway in October 1942. Initial efforts at the future townsite and plants involved large-scale grading that in turn destroyed much of the existing ground cover and made mud a perpetual problem for residents in the next years. It would become common practice for community women to arrive at formal dances on the reservation dressed immaculately except for mud-caked rubber boots which would be quickly removed and stacked in the foyer along with those of others in attendance.[15] The brownish-red substance was continually found as a layer around the bottom of project executives' trousers, and some remember it as standard etiquette to remove one's shoes before entering a house. Wooden sidewalks built and installed on the most commonly traveled routes of the townsite provided only limited

Pervasive mud symbolized early Oak Ridge (March 1943).

protection. The mud remains one of the most graphic recollections for many present day Oak Ridgers. The nature of that recollection was perhaps best expressed in the poetic efforts of one wartime Tennessee Eastman employee whose name has been lost.

In order not to check in late,
I've had to lose a lot of weight,
From swimming through a fair-sized flood
And wading through the goddam mud.

I've lost my rubbers and my shoes
Perpetually I have the blues
My spirits tumble with a thud
Because of all this goddam mud.

It's in my system so that when
I cut my finger now and then
Instead of bleeding just plain blood
Out pours a stream of goddam mud![16]

Under the final plan for the townsite, the main street—Tennessee Avenue—would run east to west roughly parallel to but approximately one block north of Highway 61 almost to the construction workers' housing area. Approximately one mile to the north and also running east to west along the top of the ridge was Outer Drive. The center of the residential community, homes and dormitories, was in between. Tennessee Avenue and Outer Drive were connected by a series of main avenues running north and south (that is, up and down the hill). Branching off both sides of the avenues were roads, circles, and lanes. When cross streets were continuous or had outlets, they were designated as circles or roads. Dead-end streets leading from the avenues were designated as lanes. In this layout a deliberate effort was made to preserve the natural contour of the land as much as possible. This procedure not only had aesthetic merit but reduced the amount of grading and other land preparation prior to construction of housing. Avenues, circles, roads, and lanes were also built to meander along the natural contours of the land.[17]

Of the cemesto houses, the most attractive and desirable
was the three-bedroom "D."

For a time MED town developers considered naming the streets of the community for past and present
Corps of Engineers' officials, but the developers rejected this notion in favor of a more systematized
organization. Beginning at the eastern end of the reservation, names of avenues would progress al-
phabetically to the west and would be given names of
states—for example, Arkansas, California, Delaware.
Roads, circles, and lanes leading from avenues were
given names beginning with the first letter of the name
of the avenue. Thus all roads and lanes from Arkansas
began with an "A," those from Delaware, a "D," and so
forth. This arrangement, the military concluded,
would simplify such matters as dwelling-maintenance
records and reporting of fires or crimes, and in general
would assist residents in finding their way about the
townsite.[18]
 While the projected size of the town increased almost
daily, the first phase of its development called for
fourteen dormitories and three apartment buildings
clustered just off Route 61 on the north side and
approximately 3,000 single-family houses dotting the
curving streets farther north and east of the highway
up to the top of the ridge. The family houses which were
built from plans obtained from the Pierce Foundation
were the heart of the community. The houses were
constructed of fiber board with cement-asbestos
bonded to each side, a material known as cemesto. They
were expected to have a life range of up to twenty-five
years, but in 1979 the houses still stand, most much
remodeled. An effort was made to space homes at
least forty feet apart, and neighborhoods would be a mix
of houses which could be drawn from among at least
ten separate designs. In accordance with Colonel James
Marshall's decision, care was taken to preserve existing trees near the houses during the construction phase.
 Within the cemesto area, homes were organized
into three fairly distinct neighborhoods. Each had its
own limited shopping facilities, four to five stores

including a grocery and a drug store. Each had its own nearby grammar school. Neighborhood One, which came to be called Pine Valley, was located on the west side of the area with its center northeast of the Midtown and Gamble Valley trailer parks. Neighborhood Two, to be designated Elm Grove, was located on the eastern side of the community with its commercial center on Tennessee Avenue. Neighborhood Three, Outer Drive was to the north, centered along the top of the ridge.

At the time the townsite was begun, plans called for a "Negro Village" that, while segregated from housing for whites, at least according to official Corps correspondence, was to be comprised of much the same type of housing provided to whites. This community was to be made up of about fifty cemesto units, four dorms, a cafeteria, a church, school, and limited stores. It was to be located north of Route 61 approximately a mile west of the Elza Gate. The proposed black community never became reality, however. Under heavy pressure for additional white housing, the Army convinced itself that a black village could wait. Conveniently, as one Corps official pointed out, blacks then obtaining employment on the project were doing so at very low occupational levels, and as such they only warranted housing in the hutment areas anyway. In a masterpiece of rationalization, this same official also added that reservation blacks seemed to prefer the low-quality hutment housing to the type that would be provided in the proposed Negro community. At any rate, housing facilities in this area were soon greatly expanded, and the location became a fourth white neighborhood known as "East Village." The military continued to talk among themselves about plans for a black community on through the fall of 1943, but it was a lost cause. Aside from regular hutment housing, in the war period blacks on the reservation were never able to obtain more than a limited number of so-called "family hutments" (structures constructed by connecting together two standard hutments).

Situated between Pine Valley and Elm Grove on Broadway Avenue was the project's principal commercial "Towncenter," later designated Jackson Square. This cluster of stores with large adjoining parking lots was designed to support a resident population of about 13,000, the estimated size of the townsite when construction began. The location was to include fifteen to twenty stores and a guest house. Purchases here and at all reservation shopping facilities would be limited to persons with a permanent resident's pass. The matter of commercial shops would provide several challenges to the military. To begin with, the Corps clearly believed in a free enterprise system, at times during the war bringing new firms to the reservation for the primary purpose of increasing competition and forcing prices down.[19] But complete realization of such a system was impossible. For one thing, security would prohibit totally free movement of firms in and out. Secondly, because commercial space was limited on the reservation and certain basic services had to be guaranteed to residents, MED officials were put in the position of ruling on what was and what was not an acceptable or needed concession. Thus, many purchasing decisions were removed by the military from the hands of on-area residents unless they went to Clinton or Knoxville.

The nature of CEW also necessitated other basic modifications in the desired private enterprise system. Because MED authorities viewed the reservation as temporary, no concessionaires were permitted to obtain a lease for anything other than for the duration of the war, a shaky thread upon which to stake an economic future. Moreover, firms invited by the Corps to enter the reservation had to make decisions about locating on the project on the most minimal information. Security precluded disclosure to prospective merchants of a projected (and later actual) size of the town or other information upon which sales volume might be estimated. Indeed, in the planning and construction stages of the town, the military itself had no accurate way to judge potential business volume. In great part for such reasons it was apparently decided early that store rent would be a percentage of profit rather that a flat monthly fee. In this manner concessionaires were partially protected against low volume and rapid market fluctuations for what, in the end, might be a temporary business venture.

Because of the semicaptive, abnormal market conditions on CEW, the military was forced to provide consumers some protection against exploitation. Avoiding the extreme of absolute price controls over and above the general price guidelines of the Office of Price Administration (OPA), the Corps built maximum profit clauses into all concession contracts. The Corps further refused to allow, under ordinary circumstances, townsite prices to exceed those of comparable goods in Knoxville.[20] Thereby, Army authorities tied reservation commercial operation to the more normal free enterprise economy in this neighboring city.

While operating a town might be a new experience for the Corps of Engineers, in the beginning at least it seemed a simple enough matter. After all, the community had been designed very carefully. Its operation had been discussed in almost endless detail. Presumably any major problems which might arise would have been anticipated and solutions prepared. What proved to be the major blight on the townsite's hoped-for blissful future was the unexpected and apparently insatiable demands of the production plants for more and more personnel. Tennessee Eastman was a case in point. In November 1942 that firm estimated its employment needs for Y-12 at 4,000. By January 1944 those needs grew to 13,500. Similar unforeseen increases in employment demands occurred at the other reservation facilities as well.

As the prerequisites of the plants rapidly multiplied so too did the necessity to expand the scope of the supporting townsite. Under these pressures the original town plan buckled and collapsed. No sooner had the first cemesto homes reached completion in the summer of 1943 than the estimate of the population for the community was increased from 13,000 to 42,000. In 1944 the town size was again revised upward to 66,000 people. The community's population finally peaked at 75,000 in the spring of 1945. Army efforts to meet these rapidly changing needs essentially took the form of a three-phase development in housing facilities.

In the first or original phase, plans had called for approximately 3,000 cemesto homes, 14 dormitories, and 3 apartment buildings. The plans also called for construction of 980 hutments and the acquisition of 1,071 trailers. Phase two was initiated late in the summer of 1943. Planning and subsequent construction resulted in the addition of 4,793 single- and multi-family houses, 55 dorms, 2,089 trailers, 527 hutment units, and 42 barracks. Between one and two hundred old farmhouses, vacated by pre-MED residents when the reservation was created, were also pressed into service as quarters. Finally, housing built in this phase consisted of several varieties of prefabricated structures, including three flattop designs provided by the Tennessee Valley Authority which were designated as A-6, B-1, and C-1 and which had one, two, and three bedrooms respectively. Set on wood posts, these flattops were built of prefabricated sections made of plywood glued to wooden frames. Heated by coal-fired space heaters, they had canvas roofs and an estimated physical life expectancy with reasonable maintenance of approximately six years. Also included were large numbers of demountable units, known as TDUs which were dismantled and transferred to the Clinton Engineer Works from projects of the Federal Public Housing Authority as far away as LaPonte, Indiana.

Above: This pre-fabricated house appears to perch on stilts.
Each pre-fab had a coal box adjacent to the wooden walk.
Below: The two-family TDU (Temporary Dwelling Unit) was
built to last only eight years.

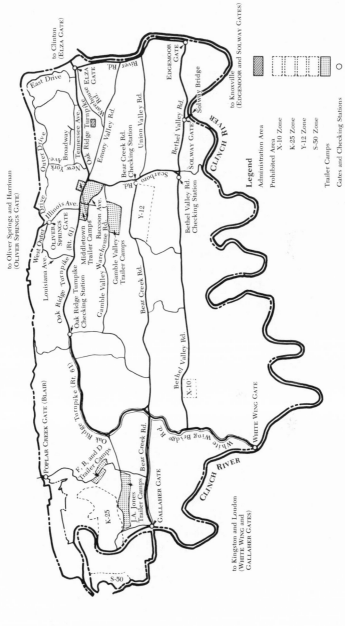

Map 2. The Oak Ridge Area

These varied extensively both in design and quality but were considered to have an average life expectancy of eight years.[21]

Until late in 1944 it was expected that no additional housing facilities would be required on the reservation. The hope was that new personnel requirements could be met through temporary expedients, such as overloading dormitories, and by very strict enforcement of criteria for on-site quarters. Plant employment needs rose steadily with each month, however, and at length it became painfully clear to the Army that further expansion in housing was necessary. Efforts were first directed toward obtaining trailers for family accommodations, but the available supply was inadequate. The final or third phase of expansion was underway by the opening of 1945. It added to the reservation 1,300 family units, 20 additional dormitories, as well as over 700 additional trailers. A substantial portion of the new family units were made up of so-called "Victory Cottages," a very poor quality structure assembled at the assigned location from prefabricated panels and parts that were chiefly constructed of plywood walls and roll roofing. The least desirable of the family units, the cottage held two one-family units, each with a single bedroom and a combination kitchen-living room. They were not expected to last over three years.

Regrettably, the exact composition of the vast population that would fill every available accommodation on or near the reservation cannot be established with any precision. For whatever reasons, but very probably those of security, CEW records contain only the most general information on occupational needs and, with limited exceptions, virtually none on job responsibilities or the educational levels of those working at the site. Reservation employment reached a peak of approximately 80,000 in May 1945. About one half of this number were classified as construction workers; 40,000 individuals were employed at the nuclear production plants; another 10,000 were associated with

Above: The Victory Cottage, offering only minor comforts, housed two families. *Below:* Foundations for pre-fabs were mass-produced.

the civilian-owned Roane-Anderson Company,
which had been chosen by the Army to manage the
townsite.[22] This peak figure does not include an unknown, though apparently not extremely large, number
of military personnel and civilian employees of the
Corps of Engineers. As the figures and general work
categories might imply, the reservation employment
population was highly diverse. Employees ranged from
a relatively small core of professional scientists, who
held graduate degrees and who possessed substantial
information about the true mission of CEW, to hun-
dreds of young women of seventeen or eighteen who
had no specialized training to operate the dials of
complex electronic equipment at Tennessee Eastman.
There were also thousands of construction workers
with varying degrees of literacy, for whom CEW was
little more than a war-related "government job."

The geographic origins of those who came to the
project are, in any detail, essentially absent from the
written record. Perhaps this omission too was a matter of
security, but more likely it is a commentary on a
pragmatic lack of interest in the subject by the Corps of
Engineers. The mission was the primary concern.
The origins of those recruited to accomplish that mission successfully was quite probably a minor concern.
What can be established is that the core of professional
scientists were largely drawn from among those in
nuclear related fields who had recently graduated or
were currently employed (many already associated
with government research efforts) at Columbia University, the University of California at Berkeley, and the
University of Chicago. Contemporary historian and
MED information officer George Robinson has
estimated that 65 percent of the total work force at the
reservation came from the states comprising the Tennessee Valley and adjoining states in the South.[23] The
majority, it would appear, came from the three states
of Alabama, Georgia, and Tennessee.

While rumors of high wages prompted many individuals to seek positions at CEW, many more people arrived at the reservation under different circumstances. In December 1943, one wartime resident recalled, "a friend and I decided to visit our husbands, both of whom worked at the Clinton Engineer Works." The telegram announcing their intentions somehow never reached its destination. Therefore the husbands did not appear at the designated point in Knoxville at 5:30 A.M. when the two arrived from Huntsville, Alabama. Determined to find their husbands and CEW, and having received what they recall as "the cold shoulder" and ill advice on directions, the two were soon lost among the narrow winding roads of East Tennessee. It was a cold, rainy, foggy morning, and they were tired after the long drive from Huntsville.

They were also becoming frightened. Ultimately, the driver halted to seek directions at a farm house on what seemed a totally desolate road. "Zeke," the mountaineer who confronted her at the door in nightshirt and cap "liked to have scared me to death. He had the most piercing eyes I had ever seen. I just burst into tears."

Contrary to appearance, "Zeke" was both sympathetic and informative, offering coffee, listening to the plight of the two women, and not only providing information on how to reach the federal reservation but also on how to locate their husbands once there. As advised, they passed through the Elza gate entry on the pretext of seeking employment at Tennessee Eastman Corporation. The mountaineer had told them that the reservation had only one dining facility (a cafeteria), and so all they need do was to wait there during the luncheon period and "sooner or later" their husbands would "turn up."

The scheme worked well in that one of the two did locate the husbands outside the cafeteria, and the couples were united. The single flaw, ultimately fortuitous, was that the second of the two women was held up in a noon-hour job interview, was offered a position with TEC, and tearfully admitted the ulterior motives involved in her reservation entrance. The offer still stood, however. She resigned from her previous job in Huntsville and became an Oak Ridger.[24]

Aside from commercial facilities, the total costs of housing development at Oak Ridge through 1946 was well in excess of $57 million.[25] Despite this expenditure and the best, indeed frantic, efforts of the Army, at no point, at least in the war period, did available on-site accommodations ever come close to demand. Moreover, by the time the townsite population hit its peak in the spring of 1945, much of the original, carefully planned residential unity and symmetry had largely disappeared. In their stead stood a grand potpourri of single and multiple family dwellings(which varied decidely in quality), large numbers of unsightly hutments, scattered and crowded trailer parks, sundry dorms of various sizes and shapes, dreary-looking barracks, and dilapidated farmhouses.

Population pressures mounted rapidly and played havoc with the Army's carefully laid plans for very limited shopping facilities, largely at Jackson Square. Because space limitations at Towncenter prohibited any significant expansion of retail services at this location and because dispersal of the rapidly growing population made it inconvenient, if not impractical, to bring all residents to the Jackson Square area even had expansion been possible, the Corps was forced to develop several additional retail centers east and west of that location. In the east near Elza Gate a shopping complex was established at East Village. Some two miles west beyond Towncenter, Middletown Center was constructed. This was a site especially designed to support the trailer camp and hutment areas. As westward residential development continued, two additional retail locations were created, Grove Center and farther to the west, Jefferson Square. Beyond these major clusters, random small concessions of diverse variety dotted the new residential areas. Fig-

Pictured here are some of Oak Ridge's bleak Victory Cottages in December 1947.

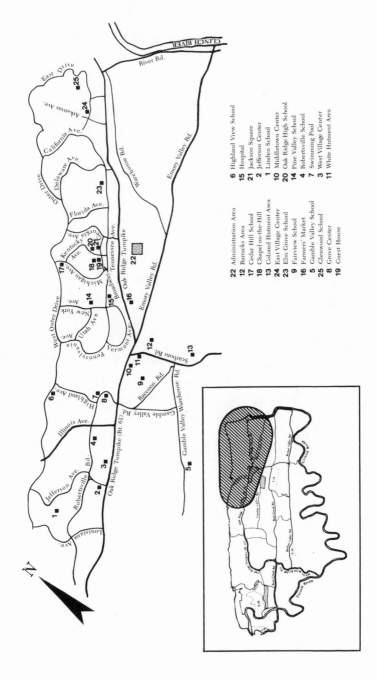

Map 3. Shopping Areas, Schools, and Centers in Oak Ridge

22 Administration Area
12 Barracks Area
17 Cedar Hill School
18 Chapel-on-the-Hill
13 Colored Hutment Area
24 East Village Center
23 Elm Grove School
9 Fairview School
16 Farmers Market
5 Gamble Valley School
25 Glenwood School
8 Grove Center
19 Guest House

6 Highland View School
15 Hospital
21 Jackson Square
2 Jefferson Center
1 Linden School
10 Middletown Center
20 Oak Ridge High School
14 Pine Valley School
4 Robertsville School
7 Swimming Pool
3 West Village Center
11 White Hutment Area

ures alone easily suffice to illustrate the demise of the original commercial vision for the project. As residents began to arrive in substantial numbers in 1943 there were only nine commercial enterprises available to meet their needs.[26] In 1945 the number of retail outlets totaled 165.

Despite the unanticipated population explosion that so severely plagued the Corps from 1943 through the end of the war, the military never lost sight of the tremendous responsibility inherent in the Oak Ridge setting where a whole community had to be built from the ground up. The town manager, Captain P.E. O'Meara, made the point well in an early issue of the reservation newspaper.

> Not since the days of our pioneering forefathers has a group of people had within their grasp an opportunity such as is offered the residents of Oak Ridge. Most of us have never before, or will never again be in a position to build our own community—to our own specifications. Most of us have never been confronted with such responsibility.[27]

Having recognized both the responsibility and opportunity involved, however, the Army never harbored utopian dreams for the community. What Corps officials did have in mind was for Oak Ridge to approximate a typical American small town as much as possible, within the constraints of security. One manifestation of this was the care taken in making the layout of the townsite attractive. Among the early steps in that process, it might be noted, was demolition of many pre-MED buildings in the area because they "were within the limits of the new townsite," and they were "unsightly."[28] Another effort toward normality, at least in part, was the decision to admit commercial stores to the reservation.

Although MED initially followed the military pattern of designating most townsite buildings and areas merely by numbers and letters, that process was reversed almost before the first structures were completed. The reservation itself was designated as the Clinton Engineer Works early in 1943, and in that summer the townsite too was given a name. Wisely rejecting a number of suggestions from MED officers in New York, the most popular of which would have doomed residents to spend their years either in Valhalla or Shangri La, Tennessee, Colonel Marshall accepted the name Oak Ridge suggested by employees on the project.[29]

In the following summer the system of street names and house numbers previously applied to houses of the town was extended to the trailer park areas. The motive was the positive "psychological effect" that would be obtained among trailer residents.[30] Somewhat later the trailer parks themselves dropped their numerical designations and received names. Again the motive was clear, a hoped-for greater sense of community among residents. In the fall of 1944 all community public facilities acquired names, for example, Arkansas Cafeteria, Jefferson Recreation Hall, Grove Center. Apparently acting with its characteristic concern for local custom, the single name submitted to and rejected by the Army was Lincoln Cafeteria. Louisiana Cafeteria, the Corps observed, would be more appropriate. The new names were also to be displayed publicly on signs which the military specified should be "as simple as possible and generally in keeping with the architectural design of the individual building."[31]

By the end of February 1945, names replaced numbers on all dorms. The names would give more identity to each building, agreed the military, and this would lead residents to take more interest in the activities of their particular dorm. In the same month the transition from military terminology in the several community groupings became complete when the primary residential shopping location, Towncenter No. 1, was officially designated Jackson Square. Further indicative of the thrust toward normalization, maps of the town showing facilities, structures, and streets both north and

The wooded hillside of Black Oak Ridge provided a small-town atmosphere for those whose jobs qualified them for a cemesto house.

south of Route 61, now known as the Oak Ridge Turnpike, became available for "general use."[32]

Among the extensive MED community records was a typed copy of a speech delivered to the Oak Ridge Kiwanis Club in late 1946 by an unidentified Corps official. The theme was an old one with the military. Security precautions must be respected, club members were told, but with regard to the town "the objective of all decisions concerning management is to approach as near as possible the normalcy of the average community of the same size."[33] By this point Oak Ridge was no longer a secret, and the town had survived the "duration of the war" time expectation built in to its creation. By then Oak Ridge had also traveled a very long way from the limited design put forward in 1942. If the original plan had been compromised, however, the spirit of it had not. The effort for normality in circumstances that were most abnormal continued, and the degree of apparent success in that effort remained a sensitive issue with the military. Taking some umbrage at the 1944 assumption of a Washington official that the townsite was essentially an Army camp, CEW spokesman Lieutenant Colonel Thomas Crenshaw promptly corrected that view in written response. "The government has built a village," he noted.[34] It was hardly an ordinary "village" but indeed Crenshaw was correct.

2. Good Fences Make Bad Neighbors

Resting beneath a forlorn marker near one boundary of the Manhattan reservation was the grave of John Hendrix, until his death in 1903 a self-appointed prophet in this area of Tennessee. Periodically he returned from trips into the deep woods with newly obtained visions. These were passed along to whomever might be willing to listen, perhaps his wife or frequently neighbors who stopped in the nearby crossroads general store. Surely his most grandiose vision was the future he predicted for the two small farming communities near his weather-beaten home, Scarboro and Robertsville. There would be a railroad in the area and the valley would be filled with great buildings and factories. Thousands of people would work in the valley, and there would be a city on the Black Oak Ridge. "I've seen it, it's coming," he would mutter and then perhaps stalk off toward the woods again. Those who paused to listen to Hendrix would often laugh at these ramblings.[1] For residents of the area in 1942 that vision might seem just as unreal, but they were no longer in a laughing mood.

When the "strangers" first showed up with their transits and survey rods, local people assumed they were simply more TVA men. But by early fall it was all too clear that this was not the case. Things fast became a nightmare. On October 6, an attorney for the Real Estate

The S-50 liquid thermal diffusion facility was not operating
in time to contribute significantly to U-235 production before
the end of the war.

Branch of the Ohio River Division, Corps of Engineers, filed in the Federal Court at Knoxville a "declaration of taking" to obtain immediate possession of 56,000 acres in Roane and Anderson counties.[2] This procedure meant that in many cases local people were forced to move off their land even before purchase arrangements were completed. Residents might well have wished they were dealing with TVA. The contrast in land acquisition policy was a sharp one. During the same period of approximately five months, October 1942 to March 1943, required by the Manhattan project essentially to conclude property transactions with those dispossessed, the Tennessee Valley Authority acquired some 52,000 acres of new land, only 3 percent of which required condemnation proceedings, while 97 percent was obtained by voluntary conveyance.[3] Residents displaced by MED action received little or no warning of what was about to occur. The informing was sudden and it was final. Not untypical was the experience recalled by one ex-resident in 1946.

> "All the folks in these parts were farmers. They worked the ground and minded their own business, peaceful folks living a simple life. We didn't pay much attention to the outside world and they didn't bother with us. That was up to 1942, anyway, when one day a man came to our house and said he was from the Government. 'We're going to buy up your land,' he said, to me. 'All of it?' I asked. 'Yes sir,' he said, 'we're going to buy all the land in this section. Everyone has to go.' "[4]

Many residents simply came home to find eviction notices tacked to the front door, a tree in the yard, or a gate. The Corps of Engineers tried to allow six weeks for evacuation but some residents had only two. The schools, churches, and crossroad groceries were closed. Homes and cemeteries were abandoned. Piling their goods on trucks or wagons, or in some cases leaving them behind, outgoing residents crossed paths with the thousands of construction workers pouring in. Many of those leaving believed that the Army had

This farmer, A.L. Robinette of the Wheat Community, was dispossessed of his property.

not given them fair value for the land. Whatever the truth of these allegations, clearly the condemnation prices did not allow purchase of comparable land holding in the suddenly inflation-ridden areas adjacent to the Manhattan project. Some of the dispossessed sought remedy in the courts and a number among them ultimately received some limited additional financial compensation. In the meantime, however, they could not stay on the land. Protests finally stirred concern in Washington. In 1943 a subcommittee of the House Military Affairs Committee investigated the land acquisition process. It came to little. The Corps of Engineers had been ordered to build an atomic bomb before the Germans did. The outcome of the war could hang on that event. Local protests, however heated, could not be allowed to slow the effort and at no point did they ever endanger project progress. "Really child's play," Corps official James Marshall later described the land acquisition process.[5]

If the military viewed this area of Tennessee as merely remote and worn-out land, residents not surprisingly saw their world quite differently. The soil might be poor and eroded, the region primitive and inhospitable, but for many in Roane and Anderson counties the land had been in their families since the nineteenth century. Others had settled there after being forced out of the Great Smokies National Park in the 1920s or out of the Norris Dam reservoir area by TVA flooding in the 1930s. A few had apparently been ejected from both the park and the Norris area and then finally also from Oak Ridge. By all accounts it was hardly an Eden-like bucolic existence. Most of the small farms had neither indoor plumbing nor electricity. The Army had to fill in almost all the wells and springs from which the people had drawn their water because they were polluted and hazardous to health.[6] But the land had belonged to them. They had lived on it, farmed it, and buried their dead in it. They would not gladly give it up. Undoubtedly speaking for many of the dispos-

Some of the farm homes evacuated were as substantial as the one shown above; other homes, like that of P.C. Stokesbury *below*, were much more modest.

sessed, one resident growled, "the only difference is when the Yankees came before, we could shoot at them."[7] Uprooted, disrupted, and embittered, many felt cheated and their hostility subsequently colored their attitudes toward the project newcomers who would become their neighbors.

To compound the problems of displaced residents, the military made no effort to assist them in their relocation. TVA land acquisition activity always included efforts to help the displaced locate new homes, but unlike the former, the Corps' concern was not to upgrade the area and the people. Their task was to get residents off the land as fast as possible, seal it from outsiders, and then build the plants and town. How evicted families moved from the reservation was their problem—and indeed it was a problem. Tires were scarce and the trucks on which they depended to remove household goods could not operate without them. In nearby Clinton, which immediately felt the brunt of the displacement, town officials even went to Nashville in an appeal to Wage and Price Stabilization authorities there for increased tire allocations. Assurances were given that any available tires would be sent to Clinton so the trucks could continue to roll.[8] But this early failure of local resources as a consequence of MED activities did not bode well for the future.

If the Manhattan dislocation represented personal tragedies for those affected, it also had enormous ramifications for the two counties in which the federal reservation was created. The least affected was Roane County to the west. In 1940 its population was slightly over 27,795. The towns of Harriman (5,620), Rockwood (3,981), and Kingston (880) made it 34.5 percent urban. Since the orientation of the project was northeast toward Clinton and east toward Knoxville rather than southwest toward Kingston and Harriman, Roane did not face the population pressures of its sister county, Anderson. In 1950 the county's population had grown to 31,665. It lost to the project, however,

Where cattle recently grazed, the foundation for the administration building was laid in December 1942.

approximately one-eighth of its land area and the tax base which that involved.[9]

Anderson County had a population in 1940 of 26,504 and a density of 78.4 persons per square mile. Although at first glance it appeared to be sparsely populated, it hid substantial numbers of people back in its hollows. The average national population density was only 44 persons per square mile. Mainly rural and agricultural, the county's largest town and seat of government, Clinton, had a 1940 population of only 2,761; the next largest, Lake City, only 1,620 people. About 15 percent of the county population lived in these "urban centers."[10]

Creation of CEW struck Anderson County with what amounted to hurricane force. The county would never be the same again. Among the first consequences was the elimination of three crossroad hamlets: Elza, Robertsville, and Scarboro which had been located on land absorbed by the federal project. A second early result was the sudden closing of Tennessee Route 61 by the MED, thus cutting the main road between Clinton and the towns of Harriman and Oliver Springs. The small county seat also became the reluctant funnel through which literally tens of thousands of workers, their baggage, and their families would be squeezed. The town of Oak Ridge at the eastern end of the reservation was located well within Anderson County about eight miles west of Clinton. As a result of population growth in Oak Ridge as well as population pressures on adjacent areas engendered by project employment needs, the county reached a peak of nearly 100,000 residents in 1945. It remained at nearly 60,000 in 1950. Approximately one-seventh of Anderson's land area was put behind the fence. This meant the loss of some $391,000 in tax revenues.[11]

In their dealings with civilians living outside the fence, the Corps of Engineers faced many unavoidable problems. Mistakes were made, however, which could have been avoided. One, as General Groves

Tennessee Avenue, hacked out of heavily wooded country, began to take shape in March 1943.

noted, was his failure to deal appropriately with Governor Prentiss Cooper. No one of sufficient rank was sent to inform the governor of the Army's plans to close off a sizeable portion of land in his state. In July 1943 when an MED representative, Captain G.B. Leonard, was finally dispatched to Nashville to tell the governor officially that Public Proclamation Number Two, which made the Clinton Engineer Works a total exclusion area, was going into effect, Cooper hit the ceiling. He had never been told anything about CEW, he growled, but after all he was only the governor of Tennessee. The Army had stolen the farmers' land and had not reimbursed the counties for expensive roads and bridges. It was, he believed, "an experiment in socialism," carried on by New Dealers cloaked under the guise of a war project. Cooper then, according to Leonard, "abruptly terminated the interview after refusing to read the proclamation and instead tearing the copies given to him in half and throwing them in the waste basket."[12] That could, and should, have been avoided. The Corps did learn from its mistakes, however. From that point forward the military acted with the greatest caution and care in all areas where reservation operation touched upon issues of state and local political sovereignty.

Since the Army and the construction and operating companies had to recruit extensively in the nearby areas for their tens of thousands of workers, it was in their best interests to maintain good relations, not only with the two levels of government, but with local Tennesseans in general. Also, since many of the workers would have to be housed off the area because of the impossibility of building sufficient on-area housing, general public hostility had to be avoided. In all this, the Army was pulled in two nearly opposite directions. It had to have the good will of the civilians surrounding the project. But in order to accomplish the removal, the secrecy, and the security, it had to take actions that alienated these same people.

Thus, relations between Oak Ridgers and outsiders began badly. With the fence and the gates and the guards there were, both literally and figuratively, insiders and outsiders. Partly because of tire and gasoline shortages, Oak Ridgers kept largely to themselves, shopping, entertaining, and making their recreation mainly on the reservation. The Army encouraged this in the interest of security. Outsiders could not get in without a purpose and a pass.

Beyond the physical separation, Oak Ridgers tended to go their own way also because in many ways they were quite different from their neighbors outside. Although over half of those working there came from Tennessee and northern Alabama, a significant minority came from far away. They were, on the average, much better educated than the people around them. Most Oak Ridgers had high school diplomas and many had college degrees. The average county resident had attended 6.8 years of school.[13] Attitudes between the two populations could and did differ substantially on many matters. Use of alcohol was an obvious example. The counties surrounding CEW were dry. It was not a condition which found much support among the often rootless construction workers as well as those from the north and west for whom prohibition had ended in 1933. Beyond that, the fact that these thousands of almost entirely young strangers had been thrown together into a kind of instant community, all working on the same highly secret, massive project, tended to establish a sense of unity and community of "us against them."

Even thirty years after the events of World War II, many Oak Ridgers still recall a definite sense that when they left the area to go to Clinton or Knoxville, they were entering hostile country. Some, when they went to shop in Knoxville, took along a pair of clean shoes so that they could not be identified as Oak Ridgers by the mud on their feet. They expected to be disliked and discriminated against in their search for goods made scarce by the war, and they frequently were. This was no doubt exacerbated by a common sense of superiority on the part of CEW residents from Chicago, New York, and Berkeley. When one of their kind married a local "hillbilly," they termed it a "mixed marriage."[14] It was significant not only as fact but as symbol that an eight-mile telephone call from Oak Ridge to Clinton was long distance.

Resentment was inevitable. In a time of national shortages, the project seemed able to get anything it wanted. Concrete evidence presented itself at every turn. Each day an average of a hundred railroad cars passed through to deposit their loads behind the fences and return empty. Oak Ridge seemed capable of swallowing up an unlimited quantity of goods of every kind; building and plumbing and electrical supplies, lumber, cement, beer, bread, and light bulbs to supply a city that grew to 75,000 people and a work force constructing plants huge beyond imagination. The observation of a local wit seemed accurate. "Tons and tons of stuff goes in and nothing ever comes out." Even if the outsiders had known about the carefully guarded grams of U-235 that were hand carried off the project, they probably would not have deemed it to be worthwhile. But they did not even know that.[15]

The apparent waste and many widely circulated stories of waste at Oak Ridge increased the resentment of those outside the fence. The case of one local who left his project job in 1944 and returned to his previous occupation selling eggs was not untypical. "Why did you resign," a customer asked? "I just quit. And I'll tell you why," the ex-employee said.

> "There they were, all those thousands of people. They were all getting good money, same as I was. And there were more buildings than you can imagine, and a lot of new roads and a lot of other new things, and they were costing a heap of money."

"I thought it all over, thought it over long and hard. And to tell you the truth, I had in mind that whatever it was the government was making over there, it would be cheaper if they went out and bought it."[16]

During World War II most Americans suffered shortages, and Clinton's and Knoxville's complaints were not uncommon. But it was easier here to transfer the unhappiness from a war that was far away, and which it would have been unpatriotic to criticize, to CEW which was close at hand and of only dubious value.

One immediate and unending pressure caused by the massive influx of people was in the area of housing. Although Oak Ridge, by mid-1945, held 75,000 people, thousands more had to be housed outside. This put an impossible strain on an area where rental housing was never especially plentiful. In Anderson County by July 1943, there were more than 650 new trailers, over half of them in Clinton, a town with approximately 600 homes. The fair grounds and the football field provided two sites for trailer camps. In the face of constant pleas for "a place to sleep—just any place," residents rented out attics and basements, garages and chicken coops. Complicating the housing situation further was the high turnover of workers. For every person who stayed, four came and left.[17]

Throughout the war, no adequate solution to the housing problem was ever found. To tap a more extensive labor and housing market, the Army inaugurated and subsidized a bus system which extended to a fifty-mile radius of Oak Ridge and became the sixth largest in the United States. By the end of 1943, an average of 20,000 people entered and left the gates every day. By that date, the Federal Housing and the National Housing Authorities had authorized 3,200 new off-area housing units, but CEW needed over 6,000. An Army study in February 1945 concluded that there was virtually no off-site housing available, and, except for Knoxville, there were no furnished rooms to be had.[18]

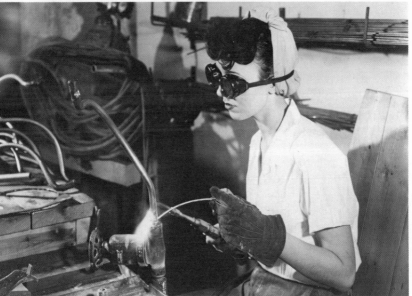

Above: Many East Tennesseans provided produce for Oak Ridgers. Here S.B. Fenn weighs beans at the Farmers Market in 1945. *Below:* Rosie the Riveter perhaps achieved more fame, but Wanda the Welder was more important at K-25 in Oak Ridge.

This saturation of housing combined with the new construction and trailers placed very heavy demands on public facilities. Clinton, which had grown from 2,761 in 1940 to 6,917 in 1943 had had no sewage disposal plant and had dumped its raw sewage into the Clinch River. With its population more than doubled and with the project only eight miles downstream, the city fathers were under very heavy pressure to build a treatment plant. General Groves added his support to the town's request for a Federal Works Agency grant-in-aid.[19]

The police force in Clinton prior to World War II had consisted of two men, one for the day and one for the night, and they provided their own transportation. They arrested one person for public drunkenness in 1940. During 1943, faced with the new workers and the prostitutes, bootleggers, gamblers, and con men who wanted to separate the workers from their high wages, an enlarged police force made more than two hundred arrests for the same offense. Even enlarging the force was difficult. The Army and the operating companies of CEW drained off the available supply of men. Roane-Anderson Company alone had a guard force of nearly 400, and the total internal security force approached 5,000. The traffic situation, wherein it became not uncommon for local drivers to be forced off the roads by reservation workers rushing to punch the clock, became so bad by November 1943 that twenty-one CEW guards were commissioned as special highway patrolmen to assist the county sheriff's department in enforcing speed limits and trying to keep order.[20]

Even the telephone system broke down. The telephone company had stated proudly in 1941 that the newly installed facilities would accommodate Clinton for at least the next four years. Less than two years later they were completely bogged down with over 4,500 calls daily.[21] The number of phones had increased 121 percent. To embitter the local taxpayers further, there seemed to be no way to get the new-comers to pay for the new facilities that their presence demanded. Most of them lived either in rental prop-erty or trailers and usually held personal possessions worth less than the constitutional $1,000 personal property exemption. A special act of the legislature would be necessary to enable the city or the county to raise money directly from these highly paid strangers, and the legislature would not meet again until Janu-ary 1945.

The seemingly endless need for skilled and un-skilled workers by the construction and operating com-panies placed demands on the local labor pool that could not be accommodated, and this need further lured vast numbers of workers away from previous jobs to the higher paying positions on the federal project. By mid-1943, common laborers on the reservation were being paid 57.5 cents per hour, a wage unprecedented in Anderson County. For the skilled trades, there was a 25 cents differential as soon as they passed through the gates onto the reservation.[22] Not surprisingly, the higher wages on the project and the flight of employees from their previous jobs left many local employers angry at CEW and suspicious of their operations. An Army intelligence officer reported a conversation with an official of the Bacon Hosiery Mills at Loudon, Tennessee, thirteen miles south of Oak Ridge. The official complained that his employees, half of whom had quit to go to Oak Ridge, came back to visit him and to say that they were doing less work for more money and that there was a lot of loafing on the job. The Bacon Mills official demanded that Senator Ken-neth McKellar force the secretary of war to explain fully what was "going on" at the project. He was appar-ently convinced that CEW might not in fact be a war project but instead, as rumor had it, some type of New Deal social experiment. Although reassured by an intelligence officer, his complaint over the loss of workers and his pleas to congressmen for help was far from unique.[23]

A very thorough study made by the Federal Security Agency in the early fall of 1944 confirmed the project-related employment problems in the surrounding area. CEW, it said, overshadowed all other producers of goods needed to support the war effort. The sixty-eight firms, exclusive of CEW, that were surveyed showed an actual decrease in the work force from 43,284 in September 1943 to 38,771 in September 1944. There was a significant need for the plexiglass, uniforms, aluminum, and other such items manufactured by these concerns, but the firms were simply unable to obtain the necessary work force to meet the demand. The work force of the Aluminum Company of America (ALCOA) had dropped from 11,671 to 8,967. Overall there was a demand for at least 5,000 more workers than could be met. The textile mills alone estimated that they needed 4,000 more workers. There was almost no construction personnel available. The FSA suggested employing more women but recognized that they only had limited job opportunities. The only hope seemed to be in the rumor that CEW was nearing completion of its construction phase and possibly would be terminating some of its employees. Other than that, the experts saw no solution.[24]

Military construction drained off machinery as well as workers nationwide. CEW compounded the problems for residents of the Anderson-Roane-Knox county area. The farmers in the Roane County Farm Bureau were unhappy in 1944 because of the Army's purchasing of farm equipment, arguing that it bought everything that was available as well as luring away farm labor. To make matters worse, the Army, to provide extra food for the residents, had begun an extensive farm program of its own in competition with local farmers. The usual protests were sent to congressmen, and, as it did when possible, the Army heeded the protests. Lieutenant Colonel Thomas Crenshaw, recognizing that the purchase of equipment and hiring of labor for the townsite grass-seeding program was upsetting the Roane County farmers, ordered the town manager not to buy any more equipment without the colonel's special permission. He did not, however, end the Corps of Engineers farm efforts, though they would be substantially cut back in 1945.[25]

A more specialized competition, and one that well represented the dilemma in which the Army was placed in its attempts to provide the good life for residents without alienating those outside the area, revolved around the beef cattle herd kept by the Corps of Engineers. East Tennessee Congressman John Jennings, responding to his constituents, complained to Secretary of War Henry L. Stimson in the summer of 1945 that the Army was slaughtering cattle in an open shed under unsanitary conditions, allowing some of it to spoil, and using bulldozers to bury valuable by-products in ditches. These animals, he said, could more economically and efficiently be slaughtered by Knoxville packing companies.

Stimson's response, drafted in Oak Ridge, spelled out the Army's difficulties. In June 1945, beef was in such short supply in the area that "the population was becoming restive." The Army feared that substantial numbers of people would pack up and leave unless something was done. Efforts to interest local packing companies in slaughtering the Oak Ridge beef herd failed because they claimed to be unable to process the cattle at the ceiling price set by the Office of Price Administration (OPA) and in accordance with the rules without a financial loss. Hence, with OPA authorization and under the supervision of three veterinarians, the Corps conducted its own slaughtering and meat-packing operation. No meat spoiled, the Army said, and all by-products were salvaged. All efforts to sell or give away the entrails, however, were unsuccessful. These were carried a safe distance from the plant, treated, and disposed of. The secretary of war hastened to assure the congressman that the policy was only to slaughter enough meat to satisfy the minimum requirements of project workers and that when adequate supplies became available the operations would be discontinued immediately. It was a reasonable approach but not one that would completely still the complaints of unfair government competition.[26]

The same issue was raised in a different form by Knoxville beer and soft drink distributors. In the spring of 1946, a local firm complained to Tennessee Senator Tom Stewart that three members of the Oak Ridge Recreation and Welfare Association (ORRWA) had formed a partnership to wholesale beer in Oak Ridge. These three had pressured breweries to deal directly with them, bypassing the Knoxville distributors. The Knoxville firm wanted its old Oak Ridge accounts back and demanded an investigation. In their usual prompt response to questions from congressmen, Corps of Engineers' officers in Oak Ridge drafted a reply to the charges which was in turn delivered to Senator Stewart by General Groves himself. He began with a summary of the formation of the ORRWA and its function of providing certain services to Oak Ridge residents prohibited to the Army by law or regulation (including the operation of beer taverns). Prior to September 1945, Groves stated, residents had depended on local beer distributors but could not get them to provide adequate quantities of popular brands like Schlitz, Pabst, and Budweiser even though they were reported to be available in Knoxville. When appeals to distributors accomplished nothing, three employees of the ORRWA formed a partnership to handle wholesale beer distribution in the area. They were given a standard concessionaires contract, but all profits from the operation would revert to the association. Those profits, some $12,000 a month, constituted a major source of revenue for ORRWA and were used to provide other services to residents such as symphonies, theater, and a newspaper.[27]

Army investigators, Groves continued, found no evidence of pressure on breweries to deprive Knoxville distributors of legitimate business or to threaten them. The partnership had, however, approached the breweries in order to get distributorships in Oak Ridge. Some had agreed; others had not. In some instances, in fact, the partnership had acted as agents for Knoxville wholesalers. An example was the very Pabst distributor who had complained to Stewart. Beer, in this instance, was ordered from Knoxville, shipped direct to Oak Ridge in railroad car lots, and unloaded and distributed by Oak Ridge employees. The Knoxville distributor received a gross profit of 33¢ per case for ordering, invoicing the association, and receiving and executing payments. While a meeting had been set up to try to iron out the complaints, Groves gave no evidence that the method of beer distribution was going to be altered.[28]

More difficult to document but clearly having a broader impact was the effect of the large numbers of Oak Ridge residents and workers who descended on stores in the surrounding communities to try to buy up available goods in a market that was already hard-hit by wartime shortages. Rumor had it that the operating companies were able to provide their employees with nearly unlimited amounts of ration stamps for meat, gasoline, and food. Workers, it was said, would arrive in stores not only with fists full of money but with their pockets stuffed with ration stamps as well. Although there was little hard evidence to support the claims of great company largess in issuing stamps, the persistent rumors revealed some deep-seated resentments.

Although the Army tried to provide adequate commercial outlets for consumer goods and services behind the fence, the demand always seemed to outstrip the supply. They were concerned that inadequate merchandise in Oak Ridge would drive workers and their families off the area, increasing absenteeism and further stretching the limited tire and gasoline supply. Hence, the Army pressed the War Production Board to increase the allotments to businesses like Miller's Department Store, whose Oak Ridge store did a gross business in 1944 of over $1,237,000.[29] Either way the decision went, however, the Army and Oak Ridgers lost in the eyes of the surrounding community. If there were not enough goods on the reservation, pressures, shortages, and local discontent increased in Clinton, Knoxville, Kingston, and Harriman. If the flow of goods to Oak Ridge increased, the resentment of the locals to "those Oak Ridgers getting everything" also grew. Beyond that, success in making the community self-sufficient and self-enclosed built the fences between Oak Ridgers and their neighbors higher and higher.

More concrete than the pressure for consumer goods in a tight market, the impact of CEW on local schools was both typical and profound. The public school enrollment in Anderson County increased by over a thousand students, from 5,228 in 1941–42 to 6,264 in 1943–44.[30] Again, the newcomers, mostly renters, provided little tax money for this increased burden. Oak Ridge, moreover, both during and after the war, paid higher salaries to its teachers, nearly double the county's salaries. The limited number of available teachers naturally gravitated to Oak Ridge. Although Knoxville raised its beginning salaries from $900 to $1,200, it still had over a hundred less-than-fully-qualified teachers on its staff. Anderson County was short twenty teachers in 1943–44. Heavily supported by federal funds, Oak Ridgers demanded, and General Groves insisted upon, an absolutely first-rate school system both in terms of teachers and facilities. If this did not become the case, employees, especially scientific personnel, would be dissatisfied, perhaps enough to leave the project. Yet if the effort succeeded, a corollary likely would be to siphon off the best teachers from surrounding communities. The reservation school system did become an excellent one but at the cost of substantial resentment over that corollary in adjacent counties. Corps officials understood the problem they had created, especially for Anderson County and Clinton. Indeed, General Groves personally urged Tennessee's Senator Kenneth McKellar to seek federal financial aid for both. Not only had they lost teachers to the project, but CEW's presence had generated enormous new enrollment pressures. Fairness, he pointed out, demanded federal help.[31] But, help or not, and whatever the local impact, the MED would do what it must to assure a superior school system for Oak Ridge. It was necessary for the successful completion of the nuclear mission which took precedence over all other considerations.

Official local hostility appeared in various forms. One of the most acrimonious was "the Battle of the Bridges" in 1943 and 1944. This matter demonstrated the complex difficulties faced by the Corps of Engineers in dealing with local government in nearby areas. The two main routes for bus and auto traffic into Oak Ridge were Solway Road and Edgemoor Road. Solway Road, nearest the plants, crossed the Clinch River at Solway Bridge. One end of the bridge was on the reservation, the other in Knox County. Edgemoor Road, coming in farther east nearer the townsite, was in Anderson County. This road crossed the Clinch River at Edgemoor Bridge and shortly thereafter entered the reservation. Although Solway Bridge was in Knox County, Anderson had contributed $27,000 toward its construction. In order to assure access to Solway Bridge and to compensate Knox County for the loss of its use (and under pressure from the governor), the Army negotiated an agreement in 1943 under which Knox County would be given $25,000 annually. The county was to spend $6,000 of that amount to maintain the southern approach road. No arrangement was made with Anderson County about Edgemoor Bridge or Edgemoor Road even though that road was closed shortly beyond the bridge.

When county officials in Clinton learned of Knox County's windfall and when the Army refused to honor their claims for $680,000 for the seventy-three miles of roads lost behind the fences, matters came to a head. A county official, Judge T.L. Seeber, told the War Department that unless Anderson County received $10,000 annually for the use of Edgemoor Bridge, he would simply order the county highway department to close it. Since the Army officers involved knew that closing the bridge would substantially delay the project and that further controversy would exacerbate what one of them described as "the general bad will we have experienced up to this time in dealing with the public around here," they bargained and settled. Anderson County got $4,000 for prior use of the bridge and $200 per month thereafter.[32]

According to Army records, Knox County did not carry out its part of the bargain to maintain the approach roads. By early 1944, the increasingly heavy traffic began to beat the roads to pieces. In February 1944, auto speed had to be cut to ten miles per hour on the two miles nearest the bridge, and after a series of bad rains the road completely failed. The Army was forced to make emergency arrangements with Stone and Webster and Knox County to allow S&W (and later Roane-Anderson Company) to maintain forty-six miles of Knox County roads. There was no way to compel the county to do the work and it had to be done. S&W provided work crews and equipment worth $5,000 per month to keep the roads passable.[33]

The bridge controversy did not affect large numbers of people who lived in Oak Ridge. Voting controversies did. It would have been surprising if those who led Anderson County and the city of Clinton politically had welcomed the project-engendered flood of potential new voters with open arms. Voting was one area where local officials, not the Army and the newcomers, were in control and the officials intended to remain so. The first evidence of this intent came in 1943 when, in the Clinton city elections, notice of the election was withheld until a week after the deadline for paying the poll tax.[34]

Because of the difficulties involved in qualifying under Tennessee state law—statutes strictly enforced by county authorities—most Oak Ridgers who participated even in the 1944 presidential election seemed to have done so by absentee ballot rather than voting locally. When Oak Ridge survived the war and it became evident to the county leadership that project residents could have considerable political consequence locally, more direct actions were attempted at times to limit that impact. In the 1945 county liquor referendum, drys stood at the polls and made lengthy challenges of the qualifications of all CEW residents. On the day before the election, a structural defect was "discovered" in the Edgemoor Bridge on the road to the polling place at Claxton school. The bridge was closed on election day and "repaired" the next day. The drys won. Two years later, in another referendum on legalized liquor sales in Anderson County, the wets won with a vote of 5,888 in favor and 4,653 against. Of the 5,888 wet votes, 5,369 came from Oak Ridge residents.[35]

The military surely desired the good will of East Tennesseans and sought actively at least to minimize feelings of hostility toward the project. Beyond this, however, efforts for community relations were minimal. Local counties and adjacent towns also made minimal efforts, an occasional welcome from the mayor of Knoxville or a dance sponsored by that city's Junior League being typical. But the hard fact was that given the reality of the CEW setting even major commitments by both would have produced limited results at best.

Demands of project security, mainly symbolized in

the fence and gates, meant that contacts between insiders and outsiders would always remain restricted. There were other "fences" as well, quite commonly significant differences in geographical, occupational, social, and educational backgrounds. Those outside identified deeply with the land, the region's past, and their own extended kinship networks. For those inside, "home" was elsewhere. They felt no sense of roots in the immediate area and generally expected to depart as soon as the war was concluded. Local residents who believed their lives disrupted and made more difficult by the existence of the project could not help but look forward to that departure. Moreover insofar as the Army succeeded in its very necessary attempt to build community spirit and a sense of unified purpose among reservation dwellers, it did so based on the uniqueness and significance of the project experience. This further drew lines between insiders and outsiders. Here, at least, good fences did not make good neighbors. In Anderson County the fences promoted a sense of separateness which would transcend the war and the opening of Oak Ridge to the general public, and that sense would still linger in diluted form even in the 1970s.

3. An Effort Will Be Made

The original design for the Clinton Engineer Works provided for direct supervision of the support community, which would become Oak Ridge, through a "town management section" of MED's Central Facilities Division. By mid-1943, however, as the projected size of that community began to mushroom far beyond the earlier expectations of the Corps, that plan seemed no longer feasible or even desirable. Supervision of the town clearly was to become a much bigger job than the Army had anticipated. There was a danger that the extensive activity which successful operation of the town would require might divert energies that should be more properly directed toward bringing the atomic mission of CEW to a speedy conclusion. Moreover, and consistent with the "normalization" concerns of the military, MED apparently determined that there were advantages in placing the community under civilian administration. Civilian control, however arbitrary it might have to become, would very likely be more acceptable to residents and employees on the reservation than direction by the Army. A civilian intermediate presence in the operation of the townsite would allow the military to maintain a much lower profile in community life than would otherwise be the case. This civilian presence could also act as a lightning rod to absorb community discontent, thus minimizing occasions for direct confrontation between reservation civilians and the Army.

The Turner Construction Company of New York City was chosen to carry the burden. The company had done work for the Corps before, most recently at the Rome, New York, Air Depot. After some negotiation Turner agreed, on a "cost plus fixed fee" basis, to relieve the Army of the details of running the town. The firm would establish for this purpose a wholly owned subsidiary created as a management firm under Tennessee law, awkwardly but appropriately labeled the Roane-Anderson Company after the two counties in which it would operate. It was born on September 23, 1943, and established offices on the project in October.[1]

The company's first project manager was Clinton N. Hernandez. He would remain in that job throughout the war years. A veteran of the first World War, Hernandez had headed his own construction firm in Yonkers, New York, for twenty-one years. Moving to Turner in February 1942, he had already carried through successfully assignments as project manager on construction jobs in Buffalo, New York, and Ashtabula, Ohio. If the Army had deliberately set out to find a proto-type representative of successful, middle-class America and the values it embodied, they could have done no better than Hernandez. He had been president of the Yonkers Chamber of Commerce, president of the Rotary Club, commander of the local American Legion post and chairman of the Board of Trustees of the First Presbyterian Church. Calling himself "a pick and shovel man," he had learned much of what he knew from experience rather than formal education. Hernandez was tough and viewed his assignment through very practical-minded eyes. "Get this straight," he would later declare in an interview, "we aren't trying any social experiments here."[2] In his person as well as in his views the new project manager epitomized the Corps' every vision of what Oak Ridge ought to be.

As intended, the Roane-Anderson Company under Hernandez's leadership moved quickly to assume its assigned duties, but the Army would retain ultimate responsibility in the community as elsewhere on the Clinton reservation. This was made clear to the residents by Captain P.E. O'Meara who retained the title of "Town Manager." In a message in the *Oak Ridge Journal* in January 1944, he spelled out the new situation. Community policy would continue to be directed by the district engineer through the Town Management Division. Roane-Anderson, as agent of the government, would simply operate reservation facilities which would otherwise demand the time of the U.S. Engineer Department. These would include the cafeterias, the dormitories, efficiency apartments, hutments, barracks, and trailer camps, the rental and maintenance of all houses, operation of the laundry and the bus system, and the negotiation and management of all concessionaire contracts. O'Meara attempted to reassure his readers by telling them that the Corps of Engineers, in establishing its policies, had given consideration to the principle of majority rule. The object, he said, even in those cases where seemingly arbitrary and inequitable decisions had been made, would be to continue to do the most good for the greatest number with the limited facilities available.[3]

On October 17, 1943, Roane-Anderson took over its first and in many ways one of its most troublesome operations—the laundry. At the end of the month, the company became responsible for custodial services of the bank, the town hall, and the post office. In November came operation of the cafeterias, the guest house cafe, some of the dormitories, the car pool, maintenance of a number of warehouses and office buildings, garbage collection, and operation of the stables which were necessary because the fence guards rode horses. The police and fire departments, although under direct military control, were paid by Roane-Anderson beginning in November. In December, the hospital personnel were added as were employees of more dormitories, the guest house, and rental houses and farm tracts.[4]

Stone and Webster, one of the major construction contractors on the site, was relieved of some of its responsibilities in November. It was informed that Roane-Anderson would operate and maintain the hospital, medical center, nurses' home, the high school, two grammar schools, the fire and guard headquarters, the telephone building, and a variety of store and office buildings. Captain E.J. Bloch of the Corps' Town Management Division urged Stone and Webster to aid Roane-Anderson by making available all the S&W personnel currently assigned to the operation and maintenance of all these structures.[5] Bus transportation, both off-area and on-area, had been very unsatisfactory and had been the cause of bitter criticism by reservation employees as well as by the Army during 1943. In February 1944, transportation was placed under the jurisdiction of a single unified Central Bus Authority controlled by Roane-Anderson. This very difficult problem had involved each major operating and construction company attempting to run free bus service for its own employees. The Army urged Stone and Webster and the other companies to transfer their bus drivers to Roane-Anderson to assure a good supply of experienced drivers.[6] As a general rule the military discouraged swapping of personnel between the various civilian companies on the project, but in such cases as the above, where one concern was permanently yielding a function to another, the Corps always actively encouraged transfer. Significantly, it should be noted that the Army could only urge, even in a situation so nearly under its full control. The military did not have the power to force the transfer of personnel employed by private firms or to force those people to remain in their jobs. The Army could only deny those who did seek transfer it thought unnecessary or unwise, continued residence in their homes if the new job was not one that qualified the employee for housing.

Because Roane-Anderson was responsible for most of the essential services provided to the community, it was the company rather than the Army which most affected the day-to-day life of town residents. The company was everyone's landlord. It delivered the coal, picked up the garbage, repaired broken windows, and tried to keep the busses running on time. The company sought out and brought necessary retail services to the reservation and was in turn held responsible by residents, as well as the Army, for the efficiency of those services. As was to be expected in a setting where rapid and virtually uncontrolled population growth combined with the fact that much of this population brought to the project pre-war civilian lifestyle expectations that could not be sustained in wartime Oak Ridge, the company received a steady stream of complaints about this, that or the other asserted failure in town operations. An MED official would later describe Roane-Anderson as "the best whipping boy the Army ever had."[7]

In retrospect it would appear that Roane-Anderson did an exceptional job in a very complex situation. Through it all, moreover, the company approached its critics and resident grievances in general with enormous patience, if mixed efficiency, in correcting specific sources of unhappiness. Good intent was always there, reflected especially in one line of prose which appeared so repeatedly in written responses from the company that the words became almost legend among reservation residents. "An effort will be made," replied Roane-Anderson. "An effort will be made."

Actually even the Army's Town Management Division (renamed Section) was not especially sanguine about Roane-Anderson's ability to carry out its manifold responsibilities and hence, particularly during the early months, looked closely over the company's shoulder. Town Manager Captain P. E. O'Meara, who also was the chief of the Operations Branch, Central Facilities Division, informed his superior, Lieutenant

Colonel J.S. Hodgson, that they might well be asking
more of Roane-Anderson than could legitimately be expected. The Operations Branch, he said, wanted
badly to turn over many of the town management functions to the company but had strong doubts about the
calibre of some of the new personnel. Usually expensive, potentially wasteful, and not especially popular
either with the Congress or the public, "cost plus" contractors were generally employed because they had
proven capabilities and were allowed wide freedom in
the direction of their work. O'Meara did not believe
that this could be the case at Oak Ridge. No company
could have had the necessary experience; this was an
entirely new job. Until Roane-Anderson gave evidence
that it could discharge its obligations, be financially
responsible, and protect the welfare of the project's general public, he recommended that Army inspectors be
assigned to monitor closely its actual operations.[8]

On the other hand, O'Meara held a very broad view
of the company's role. It was not only to perform certain
physical town management functions, but also to develop "a proper conception" of its problems. This must
include a consciousness of its responsibilities to the
public and the wisdom to preclude the possibility of
major public discontent for the sake of minor operations benefit. "Great care" had to be taken to assure that
Roane-Anderson develop a sense of mutual confidence and respect with the citizens and with the other
reservation companies. This was especially necessary since the operating contractors continued to "look
with suspicion" (a mild phrase) on such phases of the
work as the emotionally laden issue of assignment of
houses and dormitory space. Captain O'Meara correctly assumed that the Army would have to take the
position of arbiter and final authority in any dis-
agreement between different contractors or between a
single company and the public. If Roane-Anderson
did a good job, evidence of the "final authority" could be
minimized and that was a desirable goal. The military

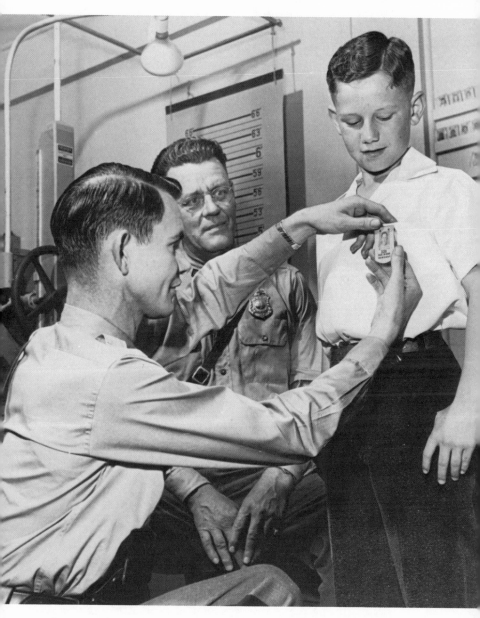

Identity badges were required for every man, woman, and child in Oak Ridge.

would make every effort to assure that a good result was obtained, including assignment to the company of a cost statistician from the Administrative Division, Cost Section, an inspector of concessionaire services, and an inspector of Roane-Anderson services. The Corps would stand constantly ready, moreover, to act directly if necessary.[9]

The Army might well be uneasy about the ability of Roane-Anderson to perform the mission assigned. That mission was enormous and manifold. Examination of the diverse responsibilities and activities assigned to the five major divisions of the company easily make the point. Within the Supervisory Division a continuing duty of the security department was to issue badges to residents, commercial operators, and suppliers. In the typical month of July 1944, however, it also conducted 146 investigations for the CEW Military Intelligence unit (which had its own operatives as well). It put up 450 posters warning of Japanese treachery and gave security education talks to some 5,000 Roane-Anderson employees. In the same month the administrative department of the Supervisory Division processed 1,852 new employees for the company, including 1,230 workers classified as non-manual, 194 skilled, and 428 unskilled. A substantial number of accidents, personal injury cases, and fires were all investigated by the safety department in that month. In the meantime the department continued its regularly assigned tasks of installing fire extinguishers, inspecting boilers and buildings for fire hazards, writing safety articles for the *Oak Ridge Journal*, and working to eliminate traffic hazards. The status of projects under construction was recorded and their cost and other relevant statistics kept by the engineer department. All forms and contracts for sub-projects, both before and after their approval by USED, were processed by this body. A second major Roane-Anderson division was utilities. The task of that unit was to assure availability of sufficient water supply and electrical power in Oak Ridge, as well as to assure the adequacy of the community's sewage system. The mission of a third division, that of accounts, is evident from its name. It carried out the routine, though very necessary jobs of time-keeping, payroll, accounts payable and receivable, finance, bonds, audit, and cost control.[10]

The mission and activities of the company's Maintenance Division were far less apparent from the name. One service which it provided was supervision of most public government transportation on the reservation. In the motor pool, Roane-Anderson supplied cars and drivers for USED, the company itself, and other CEW personnel along with Western Union and mail deliveries, convoys and off-area trips. In July 1944 ninety-seven cars carried over 21,000 passengers more than 200,000 miles.[11] Far more substantial and more fraught with difficulties for the Maintenance Division was the CEW Bus Authority. Taken over by Roane-Anderson in January 1944, that system quickly became the sixth largest bus operation in the United States. Eighteen private operators provided off-area bus service for workers at Oak Ridge from 150 communities as far away as Chattanooga, Sweetwater, Dayton, and Jellico for an average fare of 11 cents per passenger. During July 1944, these operators carried nearly 700,000 passengers a total of 1,300,000 miles in over 700 daily trips. To attract and keep the workers who lived off the reservation, fares were kept low and the system was massively subsidized. Ticket sales for July 1944 were $81,000, but the government subsidy was triple that amount—$254,000.[12]

Carrying more passengers but traveling fewer miles was the on-area service. Total mileage for the month was only about 375,000 with slightly over 1,100,000 passengers at five cents per fare. The three largest categories were townsite transportation (672,000 people), work busses to Y-12 (242,000), and to K-25 (135,000). The Bus Authority also operated free bus service for children to vacation bible schools and charter service for recreational trips to local attractions such as the parks at Big Ridge and Norris Dam. Army and Navy personnel rode free and, in the very early days, so did all other Oak Ridgers. The fare went to a nickel early in 1944 and then in August, without any advance notice to the public or consultation with the Town Council or the Central Facilities Advisory Committee, to a dime. The furor at the next town meeting accomplished two things: fares on the townsite were reduced to six cents, and the district engineer placed temporary restrictions on future meetings.[13]

Personnel in Roane-Anderson's Maintenance Division also were responsible for cleaning and maintaining all the varied housing facilities and the public and commercial buildings. Because work crews operated on a twenty-four-hour basis, meeting assignments consecutively, largely without regard to time, a certain hazard accrued to reservation residents who requested service. Stories of dinner parties, night-time slumbers, and even New Year's Eve celebrations interrupted by the arrival of Roane-Anderson employees prepared for battle with leaky faucets, stopped up toilets, or broken door locks became legendary. In retrospect amusing, these unexpected intrusions generated almost as much irritation among residents as did the lengthy wait they were often forced to suffer between the time service was requested and the time it was actually obtained. In defense of the company, some residents might have been more generous in their judgments had they realized how heavy were the demands placed upon the maintenance crews. During July 1944, a typical month, the division cleaned 544 new homes, 188 old homes from which people had moved, and washed windows in all commercial buildings and in 61 dormitories. Their electricians, plumbers, locksmiths, painters, and carpenters made 9,000 calls in East Village alone, another 7,300 in West Village and the hutment areas, and yet another 6,600 in the trailer camps. The division collected ashes and garbage in the townsite. It also distributed coal and wood. Even in hot July, 735 tons of coal were delivered to houses and apartments. An additional 700 tons was sent to other buildings as a hedge against future needs or as fuel to heat water.[14]

Beyond this type of activity, the Maintenance Division was called upon to perform an enormous amount of outdoor work. This included upkeep on 300 miles of roads within the reservation, as well as 46 miles in Knox County and 5 miles in Anderson County. The latter two categories were necessary because these roads were important avenues of approach and departure from the federal project and neither involved county would provide the necessary amount of care to keep them passable. To try to keep down the dust which was the summer's counterpart to winter's mud, the division's 25 sprinkler trucks worked daily from early morning until ten at night. Other employees were involved in spreading calcium chloride as a dust retardant. In an effort to reduce maintenance costs and improve morale of residents and workers, a program was initiated to pave reservation roads, with first priority given to those in and around the townsite. This would make heavy time demands on Roane-Anderson and Stone and Webster through most of the war years. The program only reached the roads to Y-12 and K-25 in the spring of 1945. The 163 miles of wooden sidewalks in the townsite and residential areas were either built by or kept up by the company, though the Maintenance Division was not above supporting a "do it yourself spirit" and providing boards and nails to residents who were willing to make their own repairs. Finally, the division was charged with much of the responsibility of landscape architecture in the community. This involved everything from seeding slopes to tree surgery.[15]

The heart of the Roane-Anderson Company was the Operations Division, which was charged with responsibility for every form of project housing, the laundries, the cafeterias, all contracted retail or commercial enterprises, reservation farming activities, the hospital and medical center, as well as a series of miscellaneous functions. This was the company's greatest challenge because here was the primary impact on the lifestyle of residents. As a corollary it was the division that was of greatest concern to the Army. If people were not reasonably contented with their accommodations and living conditions, they would leave the project, taking their skills, training, and experience with them. If this occurred in sufficient numbers, the mission could fail. And, indeed, even with the best efforts of the company as well as the military, four workers moved off the job during the war years for every one who stayed.[16]

Clearly the most serious problem which the company and its Operations Division inherited when it moved into Oak Ridge was housing. The essence of the problem was that throughout the war period, despite continuous and even frantic construction, the available amount of on-site housing never came close to meeting the demand for it. The housing issue was made all the more troublesome and heated because both project employers and potential employees saw on-site accommodations as a major factor in recruitment. Apparently because of the highly sensitive nature of the housing question, Corps officials would monitor this area of Roane-Anderson activity much more closely than others and would rigidly prescribe eligibility standards as well as other aspects of the housing assignment process.

Yet even here, and characteristic of Army operations in the Oak Ridge community, the military sought to avoid direct involvement with civilian residents. In the actual allocation process the Corps remained two steps removed from contact with the public. Distribution of all available forms of living accommodations were made to area operating companies, contractors, concessionaires, etc., by Roane-Anderson. Amounts of housing assigned to a given company or concern were based on percentage of employees relative to the total number of workers at CEW. These units in turn allocated the housing to their employees under guidelines previously examined and accepted on an individual basis by the Army.[17] Despite these precautions, the assignment process generated constant anger, charges of favoritism, controversy, and criticism, all largely directed at Roane-Anderson.

While the "housing issue" had a common core—the fact that supply never reached the level of demand—each form of accommodations to be assigned also posed its own particular set of problems for Roane-Anderson as well as the Army. In the case of family units, those problems were two-fold. One was a variation on the general issue of availability. The original townsite design had called for construction of approximately 3,000 cemesto single- and multi-family homes plus a very limited number of family apartments. That vision quickly collapsed before the constantly increasing personnel needs of the operating companies. As this crush mounted, the Army sought desperately to obtain any and all forms of family housing. Because of this pressure, by fall 1944, Roane-Anderson Operations found itself responsible for approximately 2,245 single-family cemesto homes, 1,973 pre-fabs, 1,447 multi-family units (duplexes and apartments), and 181 farmhouses which had been on the reservation when it was created. This was a total of nearly 6,000 family units. The number and diversity of types continued to increase almost weekly.[18]

Roane-Anderson "rode herd" on the several civilian construction companies holding contracts to erect houses, and it would accept the houses only after their completion and a careful inspection of the work performed. With this official acceptance the houses were declared available for occupancy and assigned to the appropriate companies. Neither this close supervision nor obtaining correction of building deficiencies before acceptance were easy matters. Not surprisingly, by

Four-family "E" apartments provided desirable housing for Oak Ridgers.

late 1943 and early 1944, completion of the first
thousand houses had fallen behind the projected
schedule. This brought a head-on clash between the
Army, Roane-Anderson, and the Tennessee Eastman
Corporation, since TEC had expected to receive the
largest share of these homes. Tennessee Eastman was
livid and claimed that the second phase of its Y-12
operations could not be started unless adequate personnel housing was provided. "An effort will be made,"
pleaded Roane-Anderson. Work at the plant continued.[19]

Tennessee Eastman's complaint was the failure to
deliver promised housing on time rather than the
inadequacy of the amount assigned. Carbide, the second major operating company at CEW, kept Roane-
Anderson under pressure with a slightly different complaint. Carbide feared the worst for the future because
it had a later startup date, was located much farther from
possible employment centers such as Clinton or
Knoxville, and already in January 1944 was showing a
turnover rate of 25 percent among construction workers at the plant site. K-25 would require a very large
operating force, many of whom would actually earn
less than the construction workers. Without the attraction of on-site housing, it would be impossible to
retain adequate personnel. This matter, combined with
a projected increase in the necessary work force at
K-25 from an October 1943 estimate of 500 to one of
6,000 three months later, caused Carbide to insist that
a massive new building effort be inaugurated with the
bulk of it set aside for K-25 personnel.[20]

Beyond the problems of availability, Roane-
Anderson was forced to cope with the issue of eligibility. Under reservation regulations, only bona fide
heads of families might apply for family quarters. Generally, permanent housing was restricted to those
whose salaries were over $60.00 per week base pay.
There was to be no assignment to hourly-rated employees below the classification of foreman without

79

prior approval from the district engineer's office. If employees lived within a forty-mile radius of the reservation, they would not be considered for on-area housing no matter what their pay scale or position. No houses were available for single employees.

Assignment of houses to women who were heads of households was substantially more restrictive. Even though they met all the listed criteria, which would have been difficult for most of them because of their occupational status in most companies on the area, women could not be assigned to a house or even to an apartment without the prior approval of the district engineer. Beyond the published criteria lay unwritten regulations as well. In a case that was neither unique nor isolated, in September 1944 a woman employed by Tennessee Eastman applied for a home for her family. TEC was informed that "it is not considered advisable from the standpoint of the mutual interests of the Government and Tennessee Eastman Corporation to assign a house to her or other women under similar circumstances, ipso facto. Fully explore the possibility of having another female employee share with her." Colonel Hodgson would be willing to give a woman a home if she could find another female worker to share it.[21]

Size of family governed assignment of houses; a family of two, for example, would not normally be allowed to occupy a two-or three-bedroom home. If an employee transferred from one company to another, he could usually retain his house unless he no longer qualified under "new" rules. The periodic new rules were not retroactive, however. If family size increased or if a physician advised it for medical reasons, residents could move from one type house to another if one was available. Convenience moves, such as for a better location, more shade, or more ready access to commercial facilities, were sharply discouraged and generally refused.

Because on-site housing was such a personnel attraction and because a prime concern of each civilian company was worker morale, which translated into efficient operation, every company tended to resist the rules for occupancy and repeatedly sought special concessions for its employees. In general, as the war period progressed, the direction of things was for occupancy regulations to become more firmly enforced rather than less so, and to become more restrictive rather than more relaxed. A new and tighter policy in mid-July 1944 was announced by the Army with the less-than-consoling advice: "We realize that staffing your organization will be somewhat more difficult, but this office is confident that the executives of your company will accomplish your program by directing new employees to live in trailers, or in off-site housing."[22] Conveniently for the Army, harried housing offices in these companies, under heavy pressure as they were by employees who wanted houses, took out much of their frustration on Roane-Anderson simply because that firm was charged with the job of carrying out the actual allocation process.

The problem of insufficient space to meet demand applied to reservation dormitories just as it did to houses, but administration of dorms also posed its own particular problems for Roane-Anderson Operations. With respect to the dorms, slightly more than 8,000 persons were living there by fall 1944, approximately 3,000 of whom were employed by Tennessee Eastman. The nature of the space assignment procedure varied only slightly from that employed with houses. The Corps of Engineers made space allocations and notified Roane-Anderson, along with other reservation concerns of the number of spaces available to each. The various companies informed Roane-Anderson concerning who would occupy that space. The personnel named would then be given specific room assignments by that company. Although by the war's end there were three dormitories set aside for married couples, the rest were designated for single men and women or married persons whose spouses were living away from the reservation or, as occasionally happened, living in another dormitory on-site. Commonly, with the fall of 1943, dorm rooms were no longer private. They were occupied by two to four individuals.[23]

Some dormitories were assigned completely to a single company, but more frequently personnel from various operations shared a single dorm. As was the case with houses, increasing space pressures would with time gradually force the Army to tighten eligibility restrictions and controls. At length the Corps informed Roane-Anderson that dorm assignments could be made only to operating plant employees unless prior approval had been obtained from the office of the district engineer. Applications from concessionaires, teachers, ministers, even personnel of the Roane-Anderson Company itself were increasingly rejected. Indeed, from January through April 1945, it became virtually impossible for anyone not attached to the operating plants to obtain any kind of housing on the reservation.[24]

The desperate shortage of dorm space was one of several factors which would force Roane-Anderson to involve itself in the private lives of dorm dwellers far beyond what it surely must have desired. The Army, correctly or incorrectly, suspected that individuals who had been fired, who quit or for one reason or another no longer met eligibility requirements were in fact continuing to live in the dorms. With professional personnel and shift workers coming and going at all hours and with the substantial turnover of workers which characterized the project, this was an extremely difficult matter to determine. Roane-Anderson was ordered, however, to monitor the eligibility of every dorm resident and to evict promptly those who did not qualify for room assignments.[25]

These investigations, in combination with other required intrusions on the privacy of dorm residents, constituted a set of particular administrative problems for Roane-Anderson. The problems also formed the basis of a very sore point between occupants and landlord, one that defied easy solution. In addition to inquiry into eligibility for dorm space, the company was left with the responsibility to enforce regulations among residents against gambling, possession of liquor, cooking, or visitation in rooms by members of the opposite sex. From the standpoint of Roane-Anderson and its charge by the Corps the restrictions were not unreasonable. There was clear evidence that professionals were fleecing the unwary, and gambling as well as liquor was prohibited by state law. Cooking in rooms represented a very real and significant threat of fire. Unrestricted visitation in rooms by members of the opposite sex would have offended some residents and surely would have provoked the wrath of local morality. It would have confirmed already existing suspicions of many outsiders that Oak Ridge was a place of low morals, licentiousness, and easy living.

Beyond the "image" concerns of Roane-Anderson Operations Division, the Tennessee Eastman Corporation, which employed large numbers of unmarried young women at its Y-12 operation, held a strongly paternal attitude toward them, sufficient for the company even to oppose placement of beer gardens in close proximity to dorms for women. TEC became increasingly disturbed over rumors of high-stakes gambling, drunkenness, and reports of men in women's rooms and vice versa, particularly in the dorms of the recently developed West Village area. In 1944, and under heavy pressure from TEC Roane-Anderson sought "to improve the moral aspect" of dorm life by preparing a set of conduct regulations for occupants which were to be enforced by dormitory housekeepers and clerks. If violations were serious and repeated, the guilty individual could ultimately be evicted.[26] In a further, if supremely naïve effort to upgrade life in the dorms, Roane-Anderson also sought to obtain the short-term

Above: Shown here in Oak Ridge are Jackson Square
(foreground), some of the dormitories (center), and the
administration building (upper right). *Below:* Middletown
shopping (center) provided limited goods for sale to residents
of trailer camps and hutments.

84

services of approximately ten employees from women's colleges such as Smith and Bryn Mawr to educate housekeepers and clerks in dealing with "the special problems" arising in women's dorms.[27] Dormitory residents generally responded to all such efforts by the company with reactions ranging from mild irritation to heated hostility.

Certain categories of employees were prohibited from living in reservation houses: guards, firemen, chauffeurs and chauffeurettes, bus drivers, clerical personnel earning less than $60.00 per week (without prior approval), and hourly labor (without prior approval).[28] Even if single they were also unlikely to obtain quarters in dorms. They were, however, permitted to live in trailers, barracks, and the limited number of pre-CEW farmhouses which still dotted the reservation. Here too available space never met demand, and Roane-Anderson faced administrative problems peculiar to these quarters. Basically, those problems centered on the virtually impossible task of convincing residents that their very unattractive and barely liveable accommodations were acceptable and reasonable. It was largely a losing battle. More than seven thousand trailers were crowded into the two locations of Gamble Valley and Midtown alone.[29] The trailers had no running water, and sinks drained into buckets which had to be emptied, not always promptly, by a Roane-Anderson crew. An oil stove provided the sole heat and cooking facilities. Boards beneath the stoves soon became soaked with fuel oil and had a wicked propensity to catch fire, turning a trailer to ashes before the fire department could arrive.

Farmhouses in use also generally had no running water or indoor plumbing. Most of the springs which had originally served as water supplies for pre-reservation residents had to be closed because of contamination, and it was necessary for water to be trucked in regularly by Roane-Anderson Operations. Most of the farmhouses did not have electricity, and many could be reached only by dirt roads in poor

Some of the 7,000 trailer homes in Oak Ridge had nice surroundings *above;* other trailers were located in less desirable areas.

condition. In some ways the barracks were the best of the three accommodations. The barracks, however, were spartan in appearance, minimally furnished, and made no provision even for semi-privacy; theft of resident's personal property was epidemic. Those forced to reside in the barracks did so reluctantly and had little positive to say about them.

At the very bottom of the housing heap were the hutments. Sixteen by sixteen feet square, each hut held four to six beds and had room for little else. There was a stove in the middle of each which was the single source of heat. Foot lockers or boxes for clothes and personal belongings were placed under each bed. There was no glass in the windows, only wood shutters. Central bathhouses provided the only running water, toilet, and shower facilities. Each bathhouse served twenty-four to thirty-six hutments. This form of housing was segregated by race, and among blacks by sex as well. No white women lived in hutments. Located here were the blacks and whites who stoked furnaces, made beds, cleaned floors, finished cement, dug ditches, and performed the other hard but necessary low-skilled construction and maintenance work on the reservation. The original reservation plan contemplated housing such individuals in barracks. The Army determined to employ hutments instead because the former structures in prefab form proved unavailable. Ironically, the official MED account of the history of the Clinton reservation concludes that hutment housing was superior to barracks. Hutments, states this account, had been "used with outstanding success on other construction projects," and had the advantage of promoting "improved sanitary conditions and decreased friction among workers."[30] Such a positive assessment would surely have found little support among occupants, or for that matter among other Oak Ridgers who were aware of conditions in the hutment areas.

As with the dorms, houses, and trailers, Roane-Anderson made actual hutment assignments, collected

Above: The "Colored Hutment Area" lacked scenic beauty.
Below: Hutments provided shelter from the wind and rain
but very little else.

the rents, provided such maintenance and upkeep as the areas were able to attain. The company also apparently felt that this population required special policing. Hutment sites were fenced off and regularly patrolled through the night by guards who were free to enter units at will, seeking out moral turpitude or other infractions of the rules. Because these were indeed very high crime areas, reservation police were almost constantly present there, either on patrol or in response to a call.

Housing was one of two major responsibilities held by the Operations Division of Roane-Anderson. The second was supervision of all commercial businesses and services on the reservation. The Corps of Engineers decided what goods and services should be available in Oak Ridge as well as the policy under which civilian firms would operate. Roane-Anderson then sought out appropriate concessionaires and negotiated contracts, though each contract required approval by the Army's Town Management Section before it could go into effect. Since the standard reservation contract charged concessionaires a rent set as a fixed percentage of their gross revenues instead of a flat fee, businessmen were provided some protection in the unpredictable Oak Ridge market. If sales volume dropped, rents declined. Rents themselves ranged from 1 percent to 20 percent, depending upon the type of operations. Concessions that the Army labeled to be non-essential paid the highest rate.[31]

By the time Roane-Anderson began its operations in Oak Ridge, reservation employment demands had already made obsolete the original military vision of a single community commercial center in Jackson Square, supplemented by three small neighborhood shopping locations. In 1943, however, only nine commercial enterprises, including one retail food store, opened to serve a population which by December had grown to nearly 18,000.[32] This meant that the company would have to preside over an explosion of commercial services if it were to match the burgeoning population. Because the number of retail outlets always lagged disproportionately behind population growth and because retail goods, both necessities and luxuries, were generally inadequate to demand, the company would also be a target for almost constant complaint by residents.

Nor were these the only consumer grievances with which Roane-Anderson had to deal. There were always certain trouble spots that were chronic sources of additional grief for the company. One was the laundries whose services occasioned a virtual barrage of hostile letters from residents attempting to retrieve lost items or recover compensation for damaged ones. These operations suffered from a very high labor turnover and frequent machine breakdowns. When things were going well, family bundles were processed in seven days as were cleaning orders. Dormitory and hospital bundles were normally ready in four days. Volume of goods passing through the laundries was quite large and, of course, grew as the Oak Ridge population grew. In the typical month of July 1944 they processed over 500,000 pounds of hospital and dormitory linens, took in over 16,000 individual orders, and an additional 7,664 orders in dry cleaning. Between October 1943, when these facilities were taken over by Roane-Anderson, and July 1944 the percentage of orders on which articles had been lost or damaged was 8.74 percent. Because of the total number of residents affected, this figure was quite sufficient to produce a fair amount of fist-shaking and counter-pounding at the pick-up windows, as well as to elicit the inevitable protest letters to Roane-Anderson.[33]

The cafeterias were a second major trouble spot. There were eleven, including those in the hutment areas, by July 1944. The cafeterias employed 931 persons and were serving approximately 26,300 meals per day at rates which were set at Knoxville cafeteria prices minus 10 percent. When Roane-Anderson took over the cafeterias in November 1943, they were already trouble-plagued enterprises. They were hampered by a lack of trained employees and the difficulties of acquiring a large and consistent supply of high quality food, and they were forced to meet a round-the-clock demand for service. The cafeterias seemed to have incurred the greatest popular hostility among all the operations run or supervised by Roane-Anderson. The Army began to discover the serious nature of this problem as early as fall 1943. Over a two-month period Corps officials held exit interviews with every individual who chose to terminate their employment at the reservation. One thing discovered was that in the majority of cases the true causes for departure were not those listed on the written resignation—the reasons given there were ones best suited to obtaining a clear release so that an employee could easily get other private or government employment. The interviews revealed that the greatest number departed because of the bad food and the cafeterias generally. Many of those leaving noted that they had repeatedly complained about the food but with no result.[34]

As the Corps official responsible for the exit interviews pointed out to his superiors, turnover in personnel was likely to continue to rise if something was not done to improve food service in the cafeterias. Needless to say, retention of reservation employees was a major Army concern. To that concern, as with the other types of criticisms, Roane-Anderson fell back upon its stock reply, "An effort will be made," in letters to reassure critics and complainers. Indeed a very real effort was made to minimize the causes of resident discontent. Because of the restrictions, and pressures of reservation life, as well as general scarcities imposed by the war, there were significant limitations, however, on the degree of success of any efforts made by the company. In cases such as the laundries and cafeterias the effort produced little or no change in the level of hostility.

In the meantime, with or without resident criticism, the Operations Division of Roane-Anderson had to meet its charge to develop an adequate commercial sector for the community. From August 1943, when the first drug store, grocery store, and movie theatre opened, the number and variety of concessions burgeoned. By November 1944 they included 13 barber shops, 12 grocery stores, 10 farm market areas, 8 beauty shops, 8 service stations and garages, 7 department and ready-to-wear stores, 6 shoe repair shops, and 6 restaurants and lunchrooms, as well as 3 insurance companies. The commercial sector also included such varied concerns as Fuller Brush; International Correspondence School; repair facilities for sewing machines, typewriters, and radios; a portrait studio, a tailor shop; and hobby shop. Although Corps of Engineers policy was generally to limit available space to essential goods and services, there were exceptions made from the very beginning. Contract number seven went to a flower shop and number twelve to a dancing school. Total gross receipts by these enterprises as of November 30, 1944, were slightly over eight million dollars with accrued rents of $313,000. Many more businesses would be added in 1945. Ultimately the total number of commercial establishments on the reservation would reach approximately 165.[35]

In order to assure that the government received its just share of rents and to protect residents from price gouging, the Army, through Roane-Anderson, kept a very close watch on pricing and sales practices. At times concern about rents caused the Corps and the company to respond in ways that went far beyond even those which characterized the typical profit-motivated civilian firm. So in August 1946, a full year after the close of the war, Roane-Anderson flatly refused the request of the A&P Tea Company offices in Atlanta that their Oak Ridge store be allowed to close on Wednesday afternoons as did their stores in other locations. An official of Roane-Anderson's Operations

Shoppers filled the stores, such as the A&P above, in search of
goods made scarce by war and demand.

Division reminded A&P that their contract stipulated that they would be open for business during normal business hours. If they closed on Wednesday the gross dollar volume of the store would drop and rent paid to the government would also fall.[36]

It would appear on the other hand that the military and Roane-Anderson were not always so zealous in their consumer protection efforts. A major protection practice was to peg area prices to Knoxville prices. Like other areas of the nation, Oak Ridge was, of course, subject to the general regulations of the federal Office of Price Administration (OPA) and monitoring by that agency which presumably provided further protection. Despite both however, food prices, at least, clearly did become unduly inflated from time to time. OPA investigations in the community during the summer and fall of 1944 produced what a Nashville attorney for that agency described as "atrocious violations." Eighteen owners and operators of six grocery stores were fined and enjoined from further infractions.[37]

In addition to Roane-Anderson's general responsibility for all businesses and commercial services in Oak Ridge, that firm's Operations Division, for a time, ran directly a substantial farming and cattle raising program as well. Because of normal wartime food shortages, and an evident awareness of the ill will that would be generated if the Army "raided" too heavily the agricultural resources of civilian areas surrounding the project, the military, when Roane-Anderson took over, was cultivating 640 acres of land as well as raising nearly 800 cattle and over 5,000 chickens. The new management company was expected to expand these operations further. All production would, of course, be used to meet the needs of Oak Ridge residents and employees. In 1944 another 6,000 chickens were purchased by the company, and it was anticipated that an additional seventy-three acres of land would be put into cultivation that summer. The company soon ran up against the same problem facing other farmers during the war; a shortage of labor that put limits on further expansion. The farm operation did continue, however, though in early 1945 it was significantly cut back in part, ironically, as earlier noted, because of hostility toward this activity from farmers outside the fence.[38]

Finally, and beyond the housing and commercial operations of the company, Roane-Anderson was charged with control over a number of diverse miscellaneous matters, including the supplies warehouse, the cold storage warehouse, the Western Union depot, and the hospital and medical services. The latter was by far the largest of the miscellaneous functions. With limited exceptions, physicians and dentists all held Army commissions, but Roane-Anderson's Operations Division employed and paid all other supporting personnel. Their housing, when provided, also came out of the company's quota. In a typical month by 1944 the medical service, excluding the dentists, would deal with approximately 2,100 patients, most for pre-employment physicals. The dentists cared for an additional 2,000 to 2,500 patients, and the hospital treated over 1,300 individuals.[39]

As Oak Ridge grew, the Roane-Anderson Company grew as well. When Project Manager Clinton Hernandez arrived in the fall of 1943, the company consisted of himself and three others. Because of the dramatic growth of the townsite, the company constantly fought a desperate battle to provide adequate resources which could keep the community under control. At one point in the early days in the Town Hall, sixty-five employees shared three desks and one telephone. Before the war was over there would be ten thousand people on the Roane-Anderson payroll. It began to approach that number by fall 1944. The salary inventory for the first week in November indicated a staff of 230 fireman and nearly 1,000 individuals assigned to the guard force and police. In addition, over 8,000 persons were directly employed within the five major divisions of Roane-Anderson, 3,000 of them in the

Above: Waiting lines were the despair of Oak Ridgers; shown here is one at the "clock alley" for MED headquarters. *Below:* Oak Ridgers listen to an appeal for purchase of war bonds.

Operations Division. The company's gross payroll for that week was $398,821.[40]

So the company replicated the history of the Clinton project itself. Both had grown from nothing into highly complex entities. Both would function on through 1946 and into the civilian administration of the Atomic Energy Commission. "An effort will be made," Roane-Anderson had repeatedly assured those with grievances. In their thoughtful moments most residents would probably have agreed that the "effort" was a pretty good one after all. At times of personal frustration from inadequate operations in the community, however—and these occasions were bound to happen often because of the war- and mission-imposed conditions of life on the reservation—far too many residents were less than charitable in their assessment. Perhaps it was such a passing moment of irritation which caused the response of a reservation telephone operator who refused to give out the unlisted home phone of Roane-Anderson's project manager, Clinton Hernandez. The caller sought to assure the operator that hers was not a business call. It was to extend a social invitation to the manager and his wife. "I'm one of their friends," the caller declared. "Madam," replied the operator curtly, "for your information, Roane-Anderson has no friends."[41] To the company, so often caught between the conflicting commands of the Army and the demands of reservation residents, this must at times have surely seemed the case.

In many ways the company's most profoundly ironic moment came with the close of the war. While a number of project companies received the Army-Navy E, an award for excellence in support of the war effort, Roane-Anderson was not among them. Given the magnitude of the latter's role in the successful operation of the Clinton project, this seems in retrospect a curious oversight. For at least one of the company's employees at the time it provoked much stronger feeling. In late 1945 L.A. Waddell took it upon himself to write the

district engineer for an explanation. Why other companies and not Roane-Anderson, he demanded to
know?

> We have waded in the same mud and experienced the
> same inconveniences. Roane-Anderson has enjoyed the
> dubious pleasure of being the Army's "Whipping Boy,"
> and has done a remarkable job.... While R-A bent forward
> to obey Army regulations, and orders, others applied
> the boot from the rear.[42]

There was no record of a response or explanation
amid the official correspondence and papers of the
Corps. Whatever the reason, presumably a technical
distinction between direct contribution to the war effort
versus indirect support roles, Roane-Anderson never
received the Army-Navy E. Waddell called it "a slap in
the face to a group of people who have contributed as
much toward the success of CEW as any other group on
the area."[43] The language was strong, but so was his
claim for the company.

4. Main Street by Order of...

As the townsite newspaper periodically sought to remind its readers, many problems of life in Oak Ridge were typical of all cities that had felt the effects of the war effort. Housing was scarce and, when available, often inadequate. There was rationing, and many desired commercial services were difficult to obtain. Shortages, long lines, and other inconveniences were certainly not limited to Oak Ridge. Also common was the constant stream of new faces from unknown backgrounds and the presence of sizable groups of single adults with no local family ties.

But there were very fundamental differences as well. To begin with, Oak Ridge had no core community or existing social organization when residents began to arrive in 1943. All social institutions and community structure had to be created from the ground up. Secondly, the townsite belonged to the Army. It determined who entered and left, a fact which fashioned extensively the nature of the lifestyle of reservation residents. Moreover, the Army's primary concern was, of necessity, its nuclear mission. In the end, if there were niceties of life beyond those which were absolutely necessary, or if life approached in significant degree that which was normal in most communities, the justification from the Army's point of view was that such things would keep residents happy, morale high, and job turnover low—matters that bore directly on the war mission. The mission also dictated that the military had no time to allow the townsite to grow up helter-skelter. Many social institutions and other aspects of community life were simply imposed on the town from the top down. Even those significant aspects of life which were not imposed were almost inevitably molded or directed by the nature of the project setting. Some place between the realities and necessities of that setting on the one hand, and the values and expectations of residents on the other, Oak Ridgers developed their unique style of life.

Housing was a primary case in point. Although on-site housing was all owned by the government, not unlike a military post, the diversity of that housing—ultimately a ragged mixture of single and multi-family dwellings, including well-built cemesto homes, a flat-top design taken from TVA, small prefabricated "victory cottages," dormitories, 16' x 16' one-story "hut-ments," trailers, barracks, and old farmhouses—was hardly in the military style. The basis of rent on family housing was also the opposite of the military-post model. There, rent was generally geared to salary or rank, while at CEW rent was generally related to construction costs of quarters. One advantage of this arrangement was to minimize complications and friction caused by having both military and civilians housed in the same area. It should be noted here parenthetically, however, that military men found it at least as difficult to acquire on-site family housing as did civilians. There is indeed some evidence to suggest that the Army may have received a lower priority than civilians. Certainly they were not favored.

Army personnel, especially enlisted men, experienced a varied history in terms of their ability to obtain on-site family housing, with the direction of things increasingly toward greater and greater difficulty. The question of housing for enlisted personnel was first raised with the arrival of the Military Police

Off-duty soldiers flocked to the PX for conviviality as well
as for sundries.

detachment and the much larger Special Engineering Detachment (SED). Many in the latter group were individuals who had worked on the project as civilians, been drafted, and because of their special skills and MED's high priority had been returned to continue their work as often somewhat unorthodox GI's. Initial orders issued in 1943 stated that the quarters which they and their families received would be based on the somewhat vague criteria of their "importance" to the project mission. They were also warned that there would probably be limitations on availability of houses, perhaps restricting it to senior non-commissioned officers. For those who were not married, MED tried to provide dormitory rooms. Married SED enlisted men without children would warrant apartments and, for those with children, apartments or houses.[1]

But by October 1943, regulations were already beginning to tighten up. Quarters provided for enlisted men were not to be furnished, and only the district engineer could authorize the assignment of them. Two weeks later, new orders included officers in far more sweeping restrictions for the Army. All military personnel in family quarters at that point could stay—at least temporarily—but no more would be assigned family quarters unless the district engineer determined that their duties required twenty-four-hour residence on the reservation. Military personnel were to be encouraged to move their families off the area. They were also warned that if the housing situation became more critical, those in homes might be ordered off the site on short notice. It did and they were. Early the next year MED's Colonel Marsden reported that no family quarters were available to Army personnel nor would there be in the foreseeable future.[2]

Despite the official position of the Corps, a small number of houses for officers continued to exist and, not surprisingly, they were apportioned on the basis of rank as well as duties. As in the case of civilians, the houses were assigned by the size of families, except those ranked at least lieutenant colonel were permitted to select from whatever was available. The total number of homes permanently assigned to the military in 1945 did not exceed one hundred out of the 7,400 family units, although a few additional houses were made available on a temporary basis from time to time.[3] It was also acceptable to rent rooms from families with on-site houses, and it remained possible for a military person with a dependent to obtain, if available, quarters in the few dorms for married couples provided that the dependent was employed on the project. Enlisted men could live in houses if the house were assigned to a wife who was working at CEW and who was entitled to one under her company's quota. By 1945 this concession would be further restricted so as to require that one additional worker, whether a member of the family or not, had to reside in the house "in order to obtain maximum utilization of the quarters." At one point Corps officials considered the possibility of putting enlisted men and their families in the trailers that were being vacated by construction personnel in the fall of 1944, but rejected the idea on the grounds that the personnel of the operating companies were moving in as fast as the vacancies occurred.[4]

Actually, in providing any family quarters for enlisted men or even allowing them to rent rooms from individuals renting from the government, MED was playing somewhat fast and loose with War Department regulations, and in November 1944, local officials were brought up short. Washington wanted to know "in rather complete detail" why this was being done. The justifications that came back from Oak Ridge were not particularly strong but were apparently sufficient because there was no further correspondence. The case was made that CEW authorities did not want large numbers of men in uniform living off the area for security reasons. Large numbers of uniformed personnel would call attention to the military nature of the project. By virtue of all these restrictions and dispensations, of the 1,024 enlisted men on the area in March 1945, 818 lived in barracks, 131 occupied other type quarters on the project, and 75 were living off the area.[5]

At the time the townsite reached its maximum size of approximately 75,000 residents in the summer of 1945, the general housing distribution of the entire reservation population was as follows: single and multiple family units, including apartments, supported 28,834 residents. Another 1,053 individuals resided in minimally renovated pre-project farmhouses. Townsite dorms held 13,786 residents, and the remaining 31,257 population lived in barracks, trailers, and hutments.[6] The extreme shortage of housing of any type, on or off the project, caused employees to welcome almost any accommodations offered, and this worked against the development of a class or money-oriented residential pattern, as did the construction pattern of mixing small and large houses in the same block—a practice which may have been consciously used to promote a broad sense of community. And indeed a strong sense of community did characterize Oak Ridge as a whole. It was based on a feeling of common hardship and the belief that residents were participating in something very central to the war effort, even though only a limited number knew its exact nature.

If an aim of the Army was to promote egalitarianism on the project, however, that goal was never reached. To begin with, the townsite proper was physically separated from trailer and hutment areas which largely housed the more transient, generally less educated construction workers as well as service and maintenance personnel. Certain categories of employees were, of course, precluded from obtaining houses, apartments, or even dorms and were relegated to the less favorable accommodations in trailers, barracks, and pre-CEW farmhouses. Moreover, while the record indicates a decision in 1942 to provide the same type of housing for blacks as for whites, the low occupational level of most project blacks meant that they would ordinarily warrant quarters only in the hutment areas. In addition, the Army followed the general pattern in Tennessee of racially segregated housing and facilities. Among available houses, status lines were also clearly present. The most desirable living accommodations were the well-built single-family units running along the ridge north of the townsite administrative buildings. Whatever the case in theory, in practice assignment in this area generally went to upper-echelon civilian and military personnel, a fact noted in the popular allusion to it as "Snob Hill." Indeed, residents in this location identified and spoke of a small, even more prestigious area within the larger whole labeled "Snob Nob" or "Brass Hat Circle." While the evidence does not wholly support the contention, residents believed this smaller location was retained only for highest-level military personnel. Among those assigned to houses of all types there does appear to have been a fairly sharp sense of social difference between themselves and those forced to dwell in trailers, hutments, barracks, and farmhouses. This sense of social distinction was less apparent between house dwellers and those who lived in apartments and dormitories.

That living in the best quality housing conferred or reflected a higher social status on those residents seems clear from the written record. The attitude is perhaps most plainly expressed by A.L. Baker of Kellex Corporation in a July 1944 letter to Colonel J.C. Stowers.

> Fifty-nine of our employees are in exceedingly undesirable housing, unsuited to the general type of personnel and their families who are now occupying them. We are aware of the critical need for housing and our people are fully prepared to make sacrifices required in living comfort in order to forward this project. However, certain of these people are engineers who are specialists in a particular branch of their work and their education, technical knowledge, experience and general mode of life places them and their families in a

considerably higher bracket than the average construction worker. About twenty will absolutely have to have more suitable houses. The TDU house is the most unsatisfactory and definitely not suited to the type of people we employ. Location in the Valley and generally close to very dusty areas make living conditions so particularly bad that our men are unable to persuade their families to stay and turnover is far too high. Finding replacements is almost impossible.[7]

Colonel Nichols, writing to General Groves later in the same year, agreed with the sentiments expressed in the Kellex letter and voiced his understanding that there were "higher type executive personnel" who needed special dormitory facilities with connecting baths. These facilities, he said, were very desirable in order to fill the obvious gap between ordinary dormitories and apartments or houses.[8]

Exceptions to the usually rigid rules for assigning houses and apartments could occasionally be made if the person involved were "important enough" to the project. Union Carbide, the K-25 plant operator, sought special treatment for one of its employees: "The Superintendent of Maintenance's job title is one which classes him among the first ten people of importance at Carbide. We request an exception to existing housing policy to allow Carbide to assign an E-2 furnished apartment to this man. He is married and has no children." Although E-2 apartments with their two bedrooms were to be assigned only to married couples with one child or two children of the same sex, the approval was requested as an exception and was granted by the Army.[9]

Plainly, it was possible to approach the normality of communities outside Oak Ridge in the single-family cemesto dwellings located on the hillsides north of townsite center. These homes were constructed as part of the original plan for the town or its first revisions. Here neighborhoods were laid out with less haste than in later areas and with much more care for spacing and landscape attractiveness. Residents recall the location as one where social interaction with neighbors was extensive and where children could be allowed to roam freely without concern. Occupants took pride in the upkeep of their homes and they were encouraged, through such things as ready access to tools and physical assistance in cultivation, to maintain yards. The Army, moreover, repeatedly cautioned the Roane-Anderson Company to minimize disruption of neighborhood life in carrying out necessary maintenance. In these locations the military specifically encouraged the company, though apparently with minimal success, to seek prior concurrence of tenants before arriving to perform work inside houses. The Army sought here, as in every housing area, to regulate carefully for permissible lifestyles. Activities, such as unsightly and commercial operations which appeared to lower the tone or desirability of the area, were promptly halted. The reason was clearly spelled out in a 1946 ultimatum to one resident: cease repairing cars at his home or vacate the property. Zoning regulations existed in all "progressive" towns and were designed "to make Oak Ridge a pleasant, healthy, American community in which to live."[10]

Life in the houses of CEW was, of course, not totally unaffected by the project setting. Residents felt they were forced to endure many inconveniences. High turnover of occupants in all neighborhoods encouraged a sense of instability. As project pressures mounted, residents who had somehow obtained houses that included an extra bedroom beyond actual needs were informally but steadily urged to rent them out to strangers. Under the same pressures a number of larger houses were turned into dorms for twelve to sixteen persons. Only the most fortunate of house dwellers had telephones. Residents considered themselves lucky if neighbors within walking distance had one. Doubtless telephone ownership was a mixed blessing for the fortunate, however, when they were forced to act as answering services for a number of nearby families.

Yet, by and large, life in the houses was much less troubled by the nature of the CEW setting than in other forms of accommodations. Dorms were a special problem. Those assigned to them were often the social and economic equals of individuals provided with on-site houses, and many dorm residents held the same lifestyle expectations. Such expectations were even less obtainable in dormitories than in the houses of the project. Dorms also largely housed young single people, many of whom were away from their families for the first time, or married individuals living apart from their spouses. These people were at best subject to strong periodic feelings of loneliness, and physical conditions in the dorms did not help. Quarters were generally bleak and, especially in older buildings, made all the more dismal because they were in serious need of repainting. To compound matters, by late 1943 the need for housing space became so acute that private rooms for dormitory residents were a luxury. The practice of adding additional beds and assigning two to four people to a single room was introduced.

Occupants were forced to live with a myriad of regulations which restricted their lives. While there might well be good reasons in the eyes of Roane-Anderson, residents resented being told that they could not have liquor, could not gamble, could not cook, and could not have visitors of the opposite sex in their rooms. Many dorm residents were outraged that infractions of the Roane-Anderson moral code could mean eviction from their rooms. Nor was the mood of occupants helped by the fact that their crowded living conditions meant a constant vulnerability to the series of minor but uncomfortable contagious illnesses, especially diarrhea, which periodically swept through dorms.

Yet in retrospect, many residents of the dormitories recall their lives there in nostalgic tones, especially in remembering the camaraderie of the setting. They even have little harsh to say about the ever present mud, especially pronounced in dorm areas. In the war period, however, dorm living conditions promoted substantial discontent leading to resignations as well as social problems such as alcoholism, illegitimate pregnancies, and attempted suicides. Psychological problems were severe enough by the summer of 1944 that professional counselors were placed in many dorms in an attempt to deal with them.[11] The dormitory counselors were only one aspect of the recognition that the Oak Ridge setting could pose serious psychological difficulties. While it had been hoped that such problems would not be sufficient to disrupt project operations, by the spring of 1944 it was acknowledged that this would not be the case. The primary problem, be it in houses, dorms, barracks, or hutments, was that many workers were simply not prepared to accept the limitations of existence imposed by the CEW setting. The result was establishment of a full psychiatric service within the medical facilities at Oak Ridge in March 1944.

The psychiatric service quickly turned up evidence of heavy emotional pressure very close to home. Dr. Eric Clarke, the chief psychiatrist, reported to the chief of Clinical Services that he had found alarming signs of mounting tension and physical exhaustion on the part of the hospital staff. He had recently, in his professional capacity, seen three staff members who were generally considered to be among the most competent, stable, responsible, and hard-working. These men would have been the least suspected of any inability to handle the stress. Clarke warned that the men could no longer sleep without sedatives, and they worked over seventy hours a week spread over all seven days—they had to have some relief from the pressure.

He saw the same state of physical and mental exhaustion repeatedly in patients who came to him from outside the hospital. The majority of his patients were in positions of authority and responsibility with a highly developed sense of duty and loyalty. They arrived at the hospital with a wide variety of baffling physical complaints, all of which pointed to a fundamental factor of nervous exhaustion. The only effective treatment which he could provide involved a prolonged leave of absence and freedom from pressure. Oak Ridge made heavy demands on its people.[12]

If house and dorm residents had to deal with significant psychological and physical hardships, their situation was probably envied by many reservation employees—for example, guards, firemen, bus drivers, certain clerical personnel, hourly workers,—who found themselves consigned to farmhouses, trailers, and barracks. The Army and Roane-Anderson clearly understood that there was a serious morale problem at least in the last two. In the trailer camps, an attempt was made to combat the widespread discontent by painting the trailers in colors other than the original olive drab and numbering them, and naming streets as had been earlier done in the houses and dorm areas, adding playground equipment for camp children, and refurbishing bath houses.[13] These efforts were largely unsuccessful, however, and discontent in the trailer camps remained high.

Beyond the gradual move to phase out the use of farmhouses, little was done to improve the lives of their residents. Farmhouse dwellers did have one significant advantage over trailer residents: they could use surrounding acreage to raise much of their own food.

By summer 1944, the barracks provided beds for about 780 military personnel and civilian guards, the latter employed largely by Union Carbide and Tennessee Eastman. Here the level of discontent caused serious problems from time to time in regard to security in the plant areas. There was an extreme and chronic shortage of guards. Inability of the operating companies to provide accommodations other than barracks made recruitment difficult and was a significant factor in the high turnover among guards. This apparently caused a problem sufficient in 1944 to move Union Carbide to urge that the company be allowed to use some portion of its dormitory allocation for guards. The Army granted the request.[14]

From the standpoint of psychic well-being, surely the worst living conditions on the reservation were in the hutments. As earlier described, the physical surroundings were appalling. There was no privacy and no amenities whatsoever. Hutment residents were under almost constant surveillance by Roane-Anderson guards and police who were ever ready to call them to account for rule infractions. In the case of blacks, at least, hutment occupants apparently were not permitted complete freedom to walk the streets of Oak Ridge after dark. Violence in these areas was common. Rumor had it, though incorrectly, that an average of one murder per day took place in the hutments. Unlike ex-dorm residents, there is little nostalgia among those who occupied these quarters. Remaining in memory of the time was their sense of helplessness, frustration at constant theft, and fear of physical harm. A total of 2,047 individuals lived in the white hutments and 1,481 in the "colored" hutments by the summer of 1944. The number would not change appreciably through the remaining period of the war.

As pointed out in the first chapter, the original plans for the community visualized a "Negro Village," but this became the all-white East Village. In November 1943, Colonel T.T. Crenshaw in his explanation to the district engineer provided some insights into the thinking of officers in charge of the project. By that point the first fifty houses, as well as the four dormitories and a cafeteria, were contracted for and substantially completed, and the Army had received virtually no applications from blacks for residence in the village. But because it was necessary to house a considerable number of black custodial workers, cafeteria and dormitory employees, and others whose working hours were staggered so that commuting was difficult, the Army initiated what Crenshaw said was an active campaign to persuade them to live in the village.[15]

The reasons for the failure of this campaign were "not entirely apparent to the undersigned," but he listed a number of possibilities. One contributing factor, Crenshaw felt, arose from the fact that black workers, especially women, did not adapt themselves readily to the restrictions that were necessary because of security needs on the reservation. Apparently with no irony intended, he also suggested that perhaps blacks did not want to live in the village because of the kind of housing there, "by reason of it being above the standards to which they were normally accustomed." But the main cause given by those contacted in an attempt to persuade them to live there was that they had relatives living off the reservation whom they could not leave. Whatever the causes, in the face of this lack of success in recruiting and under increasingly heavy pressure to provide housing for white employees, Crenshaw decided that the "Negro Village" could not stand vacant until such time as a more active sales campaign could provide complete occupancy. The result was East Village.[16]

The Army, of course, did only a small proportion of the hiring. Most of it was done by the private operating companies, construction contractors, and builders. While there is no evidence to suggest that these firms deliberately refrained from hiring blacks for other than very low level jobs, it seems likely that, given the general attitudes toward blacks in this period, the possibility of employing them for skilled or professional positions was simply not considered. Had blacks been employed in the latter categories and, therefore, with claim to accommodations beyond the hutments, after the "Negro Village" became East Village they could have presented the Army and Roane-Anderson with a serious problem. But this did not happen, at least in the first year. By virtue of low-level occupations alone black men and women could be relegated within current housing regulations to the hutment areas. In 1944, however, pressure began to build from within the black community in Oak Ridge for the Army to provide some type of housing for black men with families who were working in responsible jobs such as general foremen on construction crews.

The first modification in black housing came in the winter and spring of 1944. In February, Major E.J. Bloch of the Corps' Central Facilities Division authorized Stone and Webster to convert some of the hutments into family-dwelling units. This was done in a way that would keep costs to the bare minimum. Three huts were placed together in a row, or a sixteen by sixteen foot shed was built between and connecting two huts to provide a duplex unit. Each unit consisted of space in one standard hutment plus an eight by sixteen foot shed to be used as a kitchen. A door was cut into the side of the shed to allow immediate outside entrance and exit. One bath house was designated for men and one for women. In the latter, Stone and Webster removed the urinals and added a laundry sink. Each dwelling was equipped with a small coal cooking range and an ice chest. In early April, these twenty-four units were officially turned over to Roane-Anderson for allocation and supervision. The company collected the $4.00 weekly rent in advance, attempted to verify the marital status of those who applied for these accommodations, and assigned them units on a first come, first served basis rather than by specific allocation to companies as was done for the houses and dormitory space.[17] Even though there was continued heavy pressure for more changes in housing for blacks, this first modification was, "for the duration," the last.

Beyond the growing individual requests for family housing came a more forceful and formal request about the same time from a newly organized Colored Camp

Council, chaired by Robert H. Wadkins, a general foreman for Roane-Anderson. The purpose of the council
was to represent the interests and needs of the project's
black community to the Army and Roane-Anderson.
The letter which the council sent to Corps officials in
July, although very respectful, gave evidence of real
and legitimate frustration among the black residents of
Oak Ridge. First making passing mention of the abor-
tive "Negro Village," the council drew a sharp contrast
between white and black family housing. White family housing both in the apartments and in the houses had
many niceties not provided for black families: a modern kitchen, hot and cold running water, electric stoves
and refrigerators, and glass windows. The white
dwellings were also located far enough away from the
hutment areas so as to allow some degree of privacy
and home life. The black family hutments were next to
the hutments for singles, without any electrical
appliances, running water, and with wooden shutters
rather than glass windows.[18]

A meeting with the Roane-Anderson housing
superintendent had been gracious but unproductive,
and thus the council appealed to Colonel Hodgson.

> We feel that you, as a high official in the American Army
> in which so many Negro youth are fighting and dying for
> democracy and the preservation of America, will
> sympathize with the requests of those of us who are
> laboring on the home front to supply the battle front.
> We are not asking for a whole Negro town, but if some
> twenty-five or thirty homes could be set aside for
> Negroes, we would appreciate it. If that be impractical, a
> group of family huts similar to the white family huts and
> away from the labor camps would suffice.[19]

What the six signers received immediately, although
they were in all likelihood unaware of it, was a detailed
and confidential investigation of their backgrounds
and employment records. As far as can be determined
from the record, they obtained during the war years
neither the homes nor the huts.

Because of the transient nature of most of the black population and of the very close security and large numbers of guards, and because the black residents did not feel themselves bitterly abused, there were no serious racial outbreaks between blacks and whites in Oak Ridge during the war. There were minor disturbances, however, especially on the busses. A public relations study commissioned in November 1944 noted that during the preceding summer there were reports that white passengers had treated blacks roughly and complaints by blacks that bus drivers had refused on occasion to pick them up. These complaints had subsided by the time the report was submitted, partly because the bus authority had tried to schedule separate busses where there were large numbers of blacks. The so-called "Lee-Ross Study" also noted that the situation between whites and blacks had been improved as soon as signs were put up clearly designating seat arrangements that placed blacks in the back of the busses and whites in the front.

Trouble on the busses cropped up again in the spring of 1945. The police arrested "two colored boys" who had gotten off a bus and begun to throw rocks at it. That alone would probably not have been enough to cause an arrest, but it came after two or three weeks of trouble between black passengers and white bus drivers. One driver had had his arm broken, another had been beaten up, and a third had quit. Following complaints by fourteen black Roane-Anderson employees that drivers had refused to pick them up or forced them off the busses to make room for whites, an investigation ensued in which no proof of discrimination by bus drivers could be established. Drivers were subsequently flatly directed, however, that "colored passengers were to be given every consideration in conforming to the customary practice of having them move to the rear of the bus."[20]

The claim that blacks and whites were treated equally, if in the segregated manner which was basic

Above: Oak Ridgers usually provided their own
entertainment in the recreation halls. *Below:* Card playing
in the "Colored Recreation Hall" was about the only activity
open for segregated blacks.

to Tennessee during World War II, does not bear close scrutiny, at least as it relates to recreation. Blacks were not permitted to attend area movie theaters, where first-run movies were shown for an admission of 40 cents. Sitting on boxes, they might only see 16mm films at the Colored Recreation Hall for 35 cents. Although this matter was discussed at a November 1944 meeting of the Recreation and Welfare Association, apparently nothing was done. Nearly a year and a half later, in June 1946, Swep Davis, managing director of the Recreation and Welfare Association noted that the only movie service provided to residents of the Colored Hutment Area "has been two shows weekly, displayed at the auditorium of the Colored Recreation Hall."[21]

Segregation caused the other usual problems. White and black pin boys could not be employed in the same bowling alleys because of the lack of separate washrooms, and thus the Recreation and Welfare Association used white pin boys at the Central Recreation Hall and black at the other alleys. A minister of the Inter-Denominational Church complained that there were no toilet and water facilities at the West Chapel and that the nearest available facilities for worshipers were in the bus terminal. Facilities there for whites were kept locked by a private concessionaire at the time church services were held and hence could not be used. Two possibilities were that "the Army provide facilities or take the colored signs off the doors of the other toilets which are seldom if ever used by colored people so that we can use them."[22]

Religious leaders in Oak Ridge were not, however, unconcerned with the religious and moral life of the black community. A local Baptist minister expressed his conviction that more should be done to meet the needs for church services. "Much could be accomplished in the religious life of this group as a whole," he said, "if a qualified, high-type negro minister could be found."[23]

The general attitude of authorities toward blacks in

Oak Ridge during the war seems to have been accurately summed up by an official quoted in the Lee-Ross survey. "The responsibility of the Office of the District Engineer and Roane-Anderson Company is not to promote social changes, whether desirable or undesirable, but to see that the community is efficiently run and that everybody has a chance to live decently in it."[24]

Not all residents or employees were prepared, of course, to live within what the Army and Roane-Anderson considered either "decent" or permissible behavior. As in any civilian community, Oak Ridge had its share of crimes, though in fact the actual rates and the nature of those crimes tended to reflect in a significant way both the nature of the CEW setting and the population divisions within it. Understandably, in the tightly patrolled and controlled project area with exits secured by guard forces, certain forms of crime were rare, auto theft being a case in point. Similarly it was not until 1946 that an armed robbery took place in Oak Ridge. Of crimes that did take place, a disproportionate number involved individuals housed in dormitories, trailers, barracks, and hutments, with the great majority in the latter three. At the population peak in 1945 approximately 60 percent of Oak Ridge's residents lived in these accommodations, while over 85 percent of the crimes on the project were carried out by them. Available evidence suggests that the 85 percent or better figure would also be roughly applicable to crimes perpetrated by residents throughout the war period. Obviously a partial explanation for the relatively low crime rates among residents in houses would simply be the general socio-economic characteristics, but surely differing rates were related also to the more rootless nature of residents and the greater pressures of daily life in the dormitories, barracks, trailers, and hutments.

The nature of the crimes committed suggests significance in both explanations. Records indicate that well over 50 percent of all crimes investigated were alcohol-related, primarily bootlegging, drunkenness, and disorderly conduct.[25] While this area of Tennessee was legally dry except for beer sales, the infamous "Splo," as locals called their moonshine beverage, and the commercially produced liquors seemed to flow freely from countless bootleg sources. Often the liquor was brought aboard the project with great ingenuity. One common method was to conceal bottles between the skirted legs of females as they drove through the guarded gates. Another was to conceal illegal spirits under a coat in rubber hot water bottles. Some workers proved exceedingly resourceful when conventional methods failed. So it was in August 1944 that three laborers were found in a project supermarket quite drunk after having consumed large amounts of Elixir of Beef and Iron.[26] Another worker was picked up the same month for drunkenness, which he achieved by downing several bottles of a patent medicine with a high alcohol content purchased at an on-site drug store.[27] Cases of drunken behavior generally ended with a fine imposed by county authorities to whom the culprit was remanded, or with verbal censure by project police. Attempts at bootlegging, that is, bringing alcohol through the gates for resale, regardless of legal sentence, brought on almost automatic job termination.

The second most frequent category of crime, as might be expected given the environment, was disorderly conduct cases involving assault. Again, not surprisingly, the largest number of such episodes was among unskilled workers in the hutments, trailers, and barracks areas. Other common crimes were gambling and morals cases. Again, most frequent locations for such arrests, especially morals cases involving prostitution, were in the trailers and hutments. Gambling charges, however, also resulted in a substantial number of arrests in the dorms, where it was asserted periodically that professionals were at work. Nor were morals cases uncommon in dorm areas, most often involving attempts by male residents to lure maids into rooms for carnal purposes.

It would also appear that at least by 1945 the community had acquired what residents considered to be a significant juvenile delinquency problem. Its major expression came in petty crimes, fights among youthful residents, and off-hour loitering at reservation recreation halls. This was especially true of the Central Recreation Hall, where its reputation became sufficiently bad to cause some residents to stop patronizing it. The nature of the problem was more than ill-use of spare time by reservation high school students. The problem was complicated by the presence of many teenagers who had dropped out of school to take full-time jobs on the reservation and whose leisure time was dictated by the shift they worked. Project police seemed to have no effective answer to the problem nor did the Recreation and Welfare Council, which discussed it repeatedly. The Army, in December 1944, authorized the establishment of a Juvenile Court under the direction of Lieutenant P.F. Waldner, formerly affiliated with the Juvenile Court in Cleveland, Ohio. Equipped with a master's degree in social service from Western Reserve University, Lieutenant Waldner tried to work with offenders on an individual basis and within the community to prevent the vandalism and damage to government property. Another positive solution advanced by the Recreation and Welfare Association was to transform at least one of the recreation centers into a Youth Center where juvenile energies might be more closely and positively channeled.[28]

Particularly intriguing in terms of crime, especially because of the nature of Oak Ridge, were the occasional cases of vagrancy. Most such situations resulted from the failure of terminated employees to leave the area. Now and again, however, they involved cases in which vagrants had in one way or another actually penetrated the security of the project. Such was the case of one Mr. Brown picked up not once but nine times over a three-year period. Asked why he continued to return to the project after repeated arrests, he stated that "Oak Ridge had something Knoxville doesn't have." "This is high praise, perhaps," an arresting officer jotted on the record in sardonic understatement, "but area authorities would prefer that he transfer his affections to some other community."[29]

Violations of law, such as drunkenness, disorderly conduct, gambling, and illicit sexual activity, are often expressions of misused leisure time. In the summer of 1943, newly arriving Oak Ridgers began to look around for avenues of recreation. Presumably the Army clearly recognized the dangers and discontent that might be generated if appropriate avenues were not provided. As early as July of that year military officials discussed recreational needs with residents and subsequently sanctioned establishment of a civilian-run, non-profit Recreation and Welfare Association. Membership was made up of project residents, and the organization was governed by a five-man executive board which included a representative of the Army. The purpose of the association was to finance and conduct recreational programs which the Army could not fund because of legal limitations. It was anticipated that operating funds for the R&WA would largely come from revenue producing properties such as concessions, vending machines, theaters, and beer sales.

The association began by sponsoring dances in a newly completed cafeteria: by 1945 provided project residents with 2 nursery schools, 4 theaters, 6 recreation halls, 36 bowling alley units, 23 tennis courts, a swimming pool, 18 ball parks, a number of taverns, and a 9,400 volume library. Nor did this exhaust the list—the association also sponsored a variety of activities: little theater, athletic leagues, a music society, concerts, physical fitness classes, community sings, and talent shows, to list a few. In the summers the R&WA provided the community with a playground program for area children. On holidays, such as the Fourth of

July, the organization was also expected to stage parades, special dances, and competitive events. The star in the operational crown of the association was the dining hall at the Grove Recreation Hall, where as one writer for the CEW newspaper remarked, there was "expert service," "excellent food," "flowers," and "real tablecloths."[30]

In general Oak Ridge residents seemed satisfied with the recreational opportunities provided to them. Evidence does suggest, however, that facilities for blacks were never sufficient for their needs and their provision was given much lower priority than that given to white facilities. Evidence also suggests that less attention was given to the needs of hutment and trailer camp residents than to other areas. Recreational opportunities in general were enhanced, beginning in 1944, through the addition of a fairly extensive USO program, and USO personnel took a particular interest in serving the people who lived in trailers and hutments. These areas showed especially high job absentee rates, and the USO hoped to reduce them by raising morale via better recreation programs. Interestingly enough, CEW security restrictions dictated that no records or budgets of these special activities could be provided to the national headquarters and that the Oak Ridge unit could not even be listed among official USO operations across the country.[31]

Apparent unevenness of the association in recreational opportunities over the project population was also noted from time to time by the military. No, said Army officials in 1944 when the association executive board sought to increase music concert time in theaters. Additional concerts would be at the expense of movie time and would deprive Oak Ridgers who preferred movies to concerts. Moreover, if the association had extra money for concerts it might better be spent on upgrading recreational facilities in the trailer camp areas.[32] The matter of military intervention in the operation of the R&WA and in the recreational forms provided to the community is worth noting. Incidents of such intervention were common. Thus in 1944, the military demanded and obtained the termination of the Recreation and Welfare Association's business manager for allowing slot machines in the West Side Recreation Hall in violation of Tennessee law. This was an episode made all the more interesting because no one seemed sure how the machines had passed through the Oak Ridge perimeter security system. An additional illustration of Army intervention could be seen in a 1945 request by residents for a commercial broadcasting station. Absolutely not, said the security minded Army, and the matter never came up even for discussion in the R&WA.[33]

The extent of official control or influence by the military in any community activity depended greatly on the nature of that activity. At the top of that interest list was the *Oak Ridge Journal*, a CEW weekly newspaper financed by the Recreation and Welfare Association. The paper, explained an Army official, was directly related to reservation policy and public relations and therefore required more direction from the Corps. The *Journal* was begun in 1943 in the form of four mimeographed pages, and it retained that form, although enlarged in size, to March 1944 when it began to be printed commercially. By the summer of 1945 circulation was approximately 26,000 copies per issue. What "more direction" over the *Journal* by the Army actually meant for news and opinion provided to Oak Ridgers is not clear. It could and did mean that at one point the paper for security reasons was not allowed to use names of project employees in local news coverage. It meant also that in 1944 the military took direct control of the paper for a period of time despite objections from the Recreation and Welfare Association. On the other hand, the editorial staff of the *Journal* itself insisted early that, with the exception of a "few restrictions" based on security grounds, "we take orders from no one."[34]

Frances S. Gates (center), editor of the *Oak Ridge Journal*, supervises her young staff.

An area in which the Corps felt much less need to interfere, but showed a general inclination to encourage was formation of civic, patriotic, and social clubs. Such activity fitted well into the military vision of the typical American small town that the Army thought the townsite ought to approximate. Clubs were a social mainstay of such settings. Moreover, clubs were morale builders: they provided bases of cohesion and identity within a rootless and very diverse population. While theoretically retaining the absolute right to accept or reject any group, military bias toward them was sufficient to approve without comment even such a potentially controversial organization as Women for World Government. Commensurate with the Army's keenness for the small town model, however, their preference in types of clubs was quite clear. In general, local chapters of national organizations were ruled out by the Corps in the war years because of security concerns over transmitting local membership data to the National Headquarters. Apparently the single and very telling exception was approval in June 1945 to form a Junior Chamber of Commerce.[35]

Corps officials as early as the summer of 1944 were supporting, if not encouraging patriotic clubs, such as the American Legion, VFW, DAR, and fraternal groups such as the Masons, by providing them with meeting space at minimal costs—in some cases as little as one dollar per year. Oak Ridgers responded quickly to this encouragement and the possibility of grouping together in clubs. By the close of the war, the number of social, professional, patriotic, and fraternal groups exceeded thirty. This total did not include the large number of clubs formed to promote athletic activity or other groups such as the little theater organization.

If residents expected and obtained substantial recreational opportunities in their life at Oak Ridge, the large number of highly trained people employed on the project also felt strongly about the education of their children. Absence of adequate schools would create discontent, if not departure from the project. Indeed, potential residents were assured that schools would be excellent. Adequate education in the community became a major priority with the Corps. A rural Anderson County could not meet the challenge without help. The result was the characteristic blend of military and civilian resources that marked so much of life in Oak Ridge.

Beginning with three county schools located within the land that became part of the reservation, the Army built seven additional structures by 1945. Although technically the faculty was on the payroll of the Anderson County Board of Education, Oak Ridge schools were virtually autonomous. Operation and staff salaries were funded with federal money, and the whole system was operated under the supervision of an administration selected by the Army.

In an effort to fulfill the pledge to residents of a superior educational establishment, the Corps carried out a teacher recruitment effort that brought in staff from forty states. Bachelor's degrees were required for all teachers, and most high school teachers would hold master's degrees. Enticements included a pay scale significantly in excess of regular rates in Anderson County and in many cases the promise of very precious on-site housing. The result was indeed an excellent school system for resident children. From the beginning it even employed a fulltime psychologist, and by war's end it included nursery schools and kindergartens, a rarity in East Tennessee then and later. Unfortunately, blacks received only partial benefit from the system. Younger children would attend a segregated school on the reservation, but those in the high school grades would be bussed to Knoxville for their education. Nor, moreover, did whites receive the full potential of the educational complex. The limiting factor was the student parallel to the reservation's chronic over-population problem. From initial enrollment of 340 pupils in October 1943, the school population reached a crushing 8,223 in 1945. This necessitated double shifts in several elementary schools and staggered shifts in the high school, much to the distress of residents and military officials.[36]

Shelter, recreational outlets, and good education were not the only needs that had to be met if residents were to remain relatively satisfied with their existence. There were spiritual ones as well. Perhaps because of the historically sensitive issue of the separation of church and state, the Army seemed more reluctant to involve itself beyond a necessary minimum. The military vision of the townsite certainly included extensive opportunity for religious participation, however, and yet another expression of the civilian-military blend so typical of life on the reservation. Oak Ridgers were totally free to organize religious groupings, and by the end of the war there were at least twenty-two separate congregations on the reservation. Apparently the earliest such group was the United Church which was cross-denominational. Communicants merely enrolled as associate members while retaining their previous denominational affiliation. The dedication of Chapel-on-the-Hill on September 30, 1943, was as ecumenical as a World War II Norman Rockwell poster. Not only were keys to the chapel presented to representatives of various religious groups by Colonel Crenshaw, but the invocation was given by Rabbi Jerome Mark of the Jewish Temple of Knoxville; a talk was by the Reverend Eugene Hopper, rector of St. James Episcopal Church, Knoxville; the prayer for the country was offered by Father Joseph H. Siener; another talk was given by Reverend B.M. Larsen of the United Congregation; and the final prayer for those in authority was asked by Reverend R.W. Provost of Knoxville's First Baptist Church.[37] Of the twenty-two congregations, the largest came to be Catholic, followed by the Baptist and Methodist.

If congregations wanted buildings in which to wor-

The Chapel-on-the-Hill is a place of worship for those of many creeds.

ship, on-site housing for their ministers, and off-reservation ministers or even sacramental wine to pass through the perimeter gates, they had to deal with the Army. The military was quite prepared to provide meeting space for worship. This included construction of some chapel facilities, such as the popular Chapel-on-the-Hill. It also meant making space available in buildings such as cafeterias and theaters, as well as giving assistance in coordinating meeting times and locations among congregations. The military was less comfortable with requests for allocation of the almost impossible to obtain reservation housing to ministers. This was an issue which went all the way to Washington for final decision, a decision at least partly made in the affirmative because Oak Ridge authorities believed "the presence of ministers on the area would [help] abate criticism which might come to the community regarding morals."[38] Ultimately a criteria was established for such requests whereby housing was made available to ministers who could demonstrate a minimum congregation of 100, based on actual head count by Corps representatives, in attendance at services for four consecutive Sundays. As housing problems intensified in 1945, this requirement was increased to 200.[39] Such attendance counts could apparently at times be embarrassing, the official record of one such occasion in 1944 noted: "No one except the minister."[40]

Beyond these matters, the military consistently sought to leave the organization and operation of spiritual life with area residents. Thus a usually most supportive Corps promptly rejected a 1945 request by one congregation to provide the reservation with a baptistry where churches might administer the "rites of baptism by immersion." Such facilities should be the responsibility of congregations that required them, the responding officer noted.[41] Clearly, the Corps had no intention of opening itself up to the appearance of favoritism between denominations. Nor, it appears, was it willing to provide even certain impartial adjuncts to spiritual life, such as a cemetery. A 1944 request to establish a funeral home on the reservation was decided in the negative by the Army. There were sufficient facilities for embalming in surrounding towns, a Corps official noted and then added, with what at first appears curious logic, that a funeral home would probably promote desires for a cemetery.[42] Obviously, the phrasing implied, the latter was out of the question. The comment offers an interesting insight into the Army's view of the future for Oak Ridge. The facility was a temporary wartime expedient in which any promotion of a long-term commitment among residents should be avoided, and few things exceed the cemetery as a symbol of permanence.

Finally, with respect to reservation lifestyle, it should be noted that if Oak Ridgers did die, they did so in spite of an excellent health care environment. Very early plans were to rely on nearby Knoxville physicians to meet community medical needs, but this was soon abandoned. In its place, and with the assistance of the medical staff of the University of Rochester, the Army subsequently developed a very comprehensive program which provided medical and dental support and a public health component. In the summer of 1943 regular sanitary inspections of cafeterias, dorms, and other appropriate concessionaires got underway, and a small band of physicians began to service the needs of residents, services that included house calls. While other civilian cities had difficulties retaining an adequate number of doctors, Oak Ridge set a ratio of one physician per 2,000 citizens, and in spite of increasing medical demands seemingly kept them. Also in the summer of 1943, construction of a fifty-bed hospital was begun. Capacity was gradually expanded over the next years until a ratio of five hospital beds per 1,000 population was maintained throughout the war years. Interestingly enough, there was apparently no effort to establish separate hospital wings for blacks and whites. For a time, however, early in the life of the project, some blacks were taken off the area to St. Mary's Hospital and Knoxville General Hospital for treatment on the grounds that "adequate hospital facilities for colored employees at CEW were not available."[43]

Health problems treated at the medical facilities were not unusual in nature. With the exception of occasional cases where patients claimed, largely incorrectly, that they had been injured by radiation, most cases were of the type to be found in any civilian hospital. Among the first was the birth of a child six days before the hospital formally opened. Thereafter this type of business was brisk, with 6,700 births in the next seven years. The need for psychiatric services by the spring of 1944 has been noted. Given the unstable nature of community life, it might well be guessed that venereal disease was a major problem. So it was, though apparently not as great as at the Hanford, Washington, facility. While the hospital averaged 8.8 deaths per month from the fall of 1943 through 1944, these were general death rates in the community and very low ones. Even as late as 1948, Oak Ridge mortality rates were 2.7 per 1,000 as compared to 9.4 for Tennessee and slightly higher for the United States as a whole.[44] It should be noted, however, that these figures were very likely less a tribute to health care in the community than a result of the youthful nature of the reservation population.

Whatever the ailment that might afflict Oak Ridgers, they also had an opportunity to be protected by an extremely good voluntary pre-payment medical insurance plan designed to meet hospital costs (both inpatient and out-patient care) as well as medical fees. Group coverage was a desirable concession to residents, but creation of the project's own plan was forced because most customary group coverage, such as that provided by Blue Cross, could not be obtained without divulging personnel information which would violate project security. The plan went into operation in the fall of 1943 and was administered not by the Army, but rather by a non-profit Oak Ridge Health Association comprised of users. Cost of membership was low; initially $2.00 per month for single individuals and $4.00 for families. By the summer of 1945 the plan apparently enrolled about one-third of the reservation population. Because of government subsidy and the youthful nature of the reservation population, the plan actually produced a continuous profit through the war period.

As the foregoing demonstrates, the military hand was felt in virtually every significant aspect of existence at Oak Ridge. Ironically, the Army's desire for something at least approximating "normal life" on the reservation meant that the Corps consciously attempted wherever possible to reduce its own presence or interference in the residents' lives. This effort often took the form of waiving regulations in a variety of small situations involving the quality of life in Oak Ridge. Area Girl Scouts provide one example. Clearly, their presence was on the side of normality and enhanced community life, and therefore they were allowed to sell calendars on the reservation, a privilege virtually never granted any group on a one-time basis without formally obtaining a concessionaire contract. Also significant, even the Scouts were not permitted to violate area prohibitions against selling door-to-door, an effort to assure that the sleep of shift workers was not interrupted. This prohibition on solicitation was enforced well after the war, even to the point of arresting the Avon Lady. In early February 1947, a Mrs. Turner and her ten-year-old daughter were arrested in the upstairs lobby of one of the women's dormitories (Claremont Hall) for displaying Avon products and taking orders. The police informed her that she could accept orders and act as Avon's representative, but that if she continued to solicit orders or use her residence for a display and show room, they would exclude her from the area. Still another example of the attempt to pro-

Organizations such as the Girl Scouts, shown here near
X-10, provided a link with normality in wartime Oak Ridge.

133

vide what the Army considered a pleasing life-style was a 1946 decision on a food facility in Ridge Hall, which put aside government regulations against open flames in places of public assembly because, the responding officer noted, "it is recognized that the use of candle light for the evening meal ... contributes substantially to the atmosphere of the dining room."[45]

Paradoxically, the effort to promote normality could at times draw the military very deeply into most unlikely areas of resident life, even to providing residents with such diverse items as specifications for chicken coops and plans for build-it-yourself garages, including one style that might be moved from one location to another. Military response to resident complaints against other residents, however, allows perhaps the clearest insight into the Corps' views on reservation life-style. The Army consistently sought to handle these matters with a minimum of commotion and disruption to citizen freedom, a posture that at times would lead officials toward surprisingly simple and pragmatic solutions. Thus for those who complained of "beer drunks" around an area cafeteria at the dinner hour, Corps officials simply ordered the adjacent tavern to open at 7:00 P.M. rather than 5:00 P.M. When commuters on a morning bus line complained of drinking en route to work, the military promptly eliminated the stop in front of the establishment where the intoxicants were purchased. In the case of neighborhood complaints about constant fights and loud profanity by one local family, the family was simply pressured to move to a farmhouse "a considerable distance from anyone," where their highly vocal lives could continue without offending.[46]

While hardly significant in particulars, such activity was collectively rather powerful testimony on the military commitment to assure an acceptable level of life for reservation residents. Sensitivity to the quality of life in Oak Ridge, or at least the perception of it by residents, even moved the Army to commission the public relations study of this matter by a civilian firm in December 1944. In commissioning the study, however, the Corps had no illusions about totally eliminating discontent. As the firm's final report noted, many people came to Oak Ridge less for patriotic than for economic motives. They, especially, looked to a standard of service characteristic of a normal community, and, whatever the effort, this could not be reached in reservation life. Dissatisfaction was further encouraged because the town was run by persons over whom residents had no real control, and there were no competing services to which individuals could turn if existing ones proved inadequate. Such information was hardly news to Army officials. They also undoubtedly concurred with the report's assertion that it was not even desirable that residents should have a preferential liking for the town. The community was after all not one of "free independent Americans," but of "civilians subject to military rule."[47]

The townsite was part of a war project. It was not a social experiment, nor was it intended to be a perfect community. The military had repeatedly said this. Admittedly, primary motivation for the Corps' commitment to a decent quality of life on the reservation was to facilitate successful completion of the atomic mission, but the effort was no less real or earnest because of it. And, indeed, with notable exceptions such as the conditions in the hutment areas, life of residents at Oak Ridge was not unpleasant. In part this was because residents believed themselves to be in a temporary setting and seemed to be determined to live their lives as normally as the situation would allow. In part, distress over hardships of life was moderated among Oak Ridgers by their sense of the significance of their project to the war effort.

In part, also, the youthful nature of the reservation population was a factor. As one resident noted in 1945, the project was run very largely by the young. Almost all the technical men up to the executive positions were five or less years out of school. Even key men were nearly all under forty-five. The whole atmosphere of the plant operations was that of a university rather than a factory.[48] The comment was exaggerated, but assuredly the presence of so many young adults collectively in the early stages of careers and family life did encourage a sense of excitement and adventure.

In the final measure, however, the major credit for the nature of life at Oak Ridge, the good and the bad, must go to the Corps of Engineers. It was the military that first made the decision to move beyond the limited model of an Army post in developing living conditions in the townsite. It was official actions by the military that set the priorities, parameters, and possibilities of life on the reservation. It was the military that was charged with assuring that life in the townsite operated to facilitate successful conclusion of the project's overriding mission. There was no room for error and General Groves clearly understood this. "If our gadget proves to be a dud," he once observed, "I and all the principal Army officers of the project . . . will spend the rest of our lives so far back in Fort Leavenworth dungeon that they'll have to pipe sunlight in to us."[49] The prediction was overdrawn but not without its kernel of truth. In the event of failure, it would surely have been with the Army, in Harry Truman's favorite phrase, that the "buck stopped."

5. But Especially Speak No Evil

In the fall of 1942 Franklin Roosevelt officially informed General Groves that operations of the Manhattan project were to be understood as top secret and
security arrangements should reflect that fact.[1] In Oak
Ridge, as at other MED locations, this meant that the
Army had essentially three tasks. One was to restrict
physical access to the reservation, as well as certain
areas within it, to designated people. A second was to
eliminate or be prepared to meet all potential threats
to smooth operation of the project. This was a task which
meant, significantly, that virtually any occurrence
could become a security matter if it appeared disruptive.
Finally, information about the purposes and activities
of Oak Ridge must be strictly compartmentalized and
contained; even awareness of reservation personnel
would be put on a need-to-know basis. Security demands would have profound effects on conditions of
life for project residents.

One fact of substantial consequence, not merely
physically but surely psychologically as well, was the
presence of a large and very visible guard force. The
so-called "Safety Forces" came into being in February
1943, the same month in which construction began on
the electromagnetic plant and on the site of the
projected uranium reactor. Initially the Safety Forces
contained a mere two guards and four-man fire com-

Above: Almost 750 Army MPs joined some 1,300 civilian lawmen in policing Oak Ridge. *Below:* Security checks generating the most civilian resentment were at the perimeter gates.

pany.[2] In April, however, an intense security build-up
got underway. Armed guards appeared at the seven defined entry points to the reservation; fencing was put
in place at strategic points on the exposed perimeter of
the project. Mounted patrols began to move regularly
along that portion of the site bounded by the Clinch
River. By early 1945, visible security forces included
4,900 civilian guards, 740 military policemen assigned to three detachments, and more than 400 civilian
policemen.[3]

Organizationally, each civilian operating plant provided its own security force. Roane-Anderson guards
protected the administrative area as well as manned
the four gates which provided access to that area and to
the community. The military controlled three gates
providing entry into the so-called "prohibited" plant
areas, covered the perimeter of the reservation, and
carried on roving patrols throughout the reservation.
Civilian police units were specifically charged with
traffic control and protection of the Oak Ridge community, including all housing locations.

Unlike the normal American community, Oak Ridgers
constantly lived against a backdrop of uniformed and
suspicious agents of military or civilian authority. They
were everywhere. The contrast to normal cities is
easily demonstrated in the case of civilian police. While
other southern communities of comparable size in
1945 generally had a ratio of about 1.6 police per 1,000
population, Oak Ridge had a ratio of approximately
14 per 1,000.[4] Yet numbers did not tell the whole story.
Residents, workers living off the area, and visitors
were tightly restricted in where they might go on the
reservation by an elaborate system of coded badges.
Roving military patrols periodically established random
temporary road blocks to assure that those stopped
were where they should be. Individuals passing from
point to point, even in portions of the community
area, were repeatedly called upon to present identification, and every occasion which involved police, be it

lost child or argument among neighbors, were occasions for security checks. When entering or leaving the project, proper identification also had to be presented to guards.

An additional source of possible frustration to residents, and to the military as well, was that so many members of the police and guard force were ill-experienced and limited in capability. The general manpower shortage faced by the project as a whole was further compounded in the case of these two groups because their wages were significantly less than might be obtained by persons with similar qualifications from other reservation employers. This meant that there was a constant and high rate of turnover and that many of those who did remain were limited in talents. The minimum education requirement for guards was only completion of the sixth grade, and the actual educational level tended toward that minimum. While documentation is insufficient to establish an overall generalization, it is nonetheless noteworthy that among a substantial number of Roane-Anderson guards who were given "mental ability tests" in 1945 the average mental age obtained was slightly under twelve years of age.[5]

Less a matter of resident concern, although the difficulty of dealing with the security forces was increased, was that not even the Army, much less the police and guard force, knew exactly what their power was. Obviously, they could enforce project regulations, but because the state of Tennessee had successfully retained legal jurisdiction over the reservation, guards and police had only such arresting authority as did any other ordinary citizen of the state. Unless deputized by Roane or Anderson counties, as many in time would be, they were not authorized to execute a criminal warrant or any civil process. Persons deemed guilty of criminal or civil offenses by area authorities had to be handed over to county officials for legal action.

Ill ease among MED authorities about the legal status of guards finally caused them to designate these individuals as auxiliary military police of the Army's Fourth Service Command. While this designation did not totally clarify their status, it did provide the Army with unquestioned authority to arm the guards and to obtain direct and absolute jurisdiction over all project guards in the event of emergency. The designation also acted as a deterrent to labor union activity within the guard force and thus possibly forestalled disastrous strikes by guards.[6]

When arrests were made by guards or police, the decision concerning how infractions would be handled was determined by a review board established by the Corps' Oak Ridge Intelligence and Security Division. The board operated twenty-four hours a day and seven days a week. It examined evidence in all arrests to establish, first of all, whether offenders should be handed over to county police or, on occasion, to intelligence forces. If infractions involved area regulations, the board also determined what administrative actions should be taken. While the board could not impose fines, it could recommend to appropriate project offices measures from official reprimand to job termination, or even loss of reservation housing.

Recommendation for job termination was a particularly severe matter for a reservation resident. It not only meant economic loss, but because housing and justification to be in Oak Ridge were absolutely dependent on employment, such an individual lost his home as well as his right to remain on the area. Job termination generally also meant an absolute end to employment at the project. There were few second chances. While the military refused to admit it, discharged employees were commonly placed on a blacklist whereby they became ineligible for any future position on the reservation. Roane-Anderson gave "Off-Area Releases" to persons they terminated, which meant that they were not to be re-hired by Roane-

Anderson. In addition, their names and occupations were passed on to the Army's Labor Relations Section.[7] Indeed, it would appear that individuals so listed could not even obtain visitor passes.[8]

The matter of passes—or, more properly, who came on the reservation—was an important one. Clearly, the greatest single place where security demands generated the most civilian resentment and even confrontation was at the perimeter gates. Problems took several forms. One was the continued attempt by curious Tennesseans from nearby areas to enter the reservation and "look around." These episodes were especially common on busses from Knoxville, and such individuals often did not take rejection easily. Another occasion for discontent resulted from the delay and inconvenience residents and commuters experienced in passing through the gates. Since entrance or exit required presentation of proper identification, this procedure slowed traffic and, if for some reason identification was not in order, could take a frustrating amount of time. On at least one occasion that frustration reached the point of a scuffle with guards in which the offending individual was actually shot to death.[9] Guards were also required to search a certain percentage of automobiles. Because that number was at least 3 percent even in peak hours when over 1,000 per hour were attempting to pass through some reservation gates, an inordinate delay could result.

Finally, there was the "red tape" experienced by residents and concessionaires in trying to obtain clearance for visitors. It took time and often a great deal of patience. Moreover it was never quite certain if one's efforts would prove successful. Whether commercial entertainment groups and athletic teams would be allowed to enter provided examples. There was no consistency even about who might set up outside the gates. Thus in July 1944, Army officials rejected a Recreation and Welfare Association request to allow operation of a carnival just beyond the Elza Gate despite

the fact that carnivals had been allowed to locate
there previously.[10]

There were those who perceived some advantages
in living beyond the gate. Witness the case of one young
divorcee who sought to enter the reservation in an
effort to obtain custody of a child currently living with
her ex-husband, a project resident. Because passes
could only be obtained by request of someone affiliated
with the project and her ex-husband refused to make
the request, this effort failed.[11] Indeed a common recollection among some wartime residents—largely
housed in apartments or houses—was that the gates
provided a sense of protection and safety. Even the
Corps apparently felt that reservation security, the gates
in particular, made some difference here. Within
three months following the end of the war, the Army
expressed fears to Roane-Anderson that concessionaires had developed a "false sense of security" as a
result of military presence. They should be warned
that greater individual protection and safety measures
would be necessary as military regulations were
gradually relaxed.[12]

But any advantages of the gates were clearly out-
weighed by the disadvantages for most Oak Ridgers—
and the military knew it. Discontent of residents actually became sufficiently strong to prompt formal consideration in 1944 of lifting security restrictions, including gate control to the townsite. To do so would
presumably raise morale of residents and would also
be consistent with overall military hopes to make life in
the community as normal as possible. Ultimately this
degree of "normalization" was rejected by the Army,
apparently on two grounds. One was security; every
individual who entered the area, even the townsite, was
exposed in varying degree to "classified information"
and therefore the number receiving passes had to be
limited and selective. The second, interestingly
enough, was related to the type of people who might be
attracted to an open townsite. As one officer put it,

143

open gates would bring an "influx of such parasites as prostitutes, confidence men, and thieves, as well as all sorts of preachers, social reformers, evangelists, and crackpots." Such a flow might cause more dissatisfaction among area residents than did the existing situation.[13] Concessions were made to residents, however. They would henceforth be allowed to enter the reservation through any gate, whereas entry had previously been confined to the Elza and Oliver Springs gates. Visitor passes became easier to obtain. Partly in an effort to raise the morale and status of residents and partly in the hope of speeding up incoming traffic through the gates, special "no-search" passes were provided to town dwellers, which allowed them to move through the gates without search of their car or person, for a time in 1944. Resentment and charges of discriminatory treatment by other project employees, soon forced an end to this practice.[14] Search procedures at the gates most commonly used by those who lived on the reservation, however, did become more relaxed.

Security concessions on entry to the townsite were not the only ones made to residents in 1944. The military desire to make community life as normal as protection of the project mission would allow, plus efforts to raise morale among residents, prompted the Army to introduce a political role for inhabitants. It was a limited one, only advisory in nature, but nonetheless a significant one. As Corps officials surely must have recognized, once that component was created, refusal to respect its views—or a later decision to abolish it—could produce major morale problems. One part of this civilian dimension was regularly scheduled town meetings beginning in February 1944, and the subsequent creation of an advisory Town Council. Initially members of the council were nominated and elected from the floor at the town meetings, but later in the year the townsite was divided into districts from which representatives were selected, based on popular vote of residents. The idea was that town meetings would provide a general sounding board for public opinion. The council would in turn channel that opinion to Corps officials along with its own petitions and policy recommendations.

The fact is, however, that neither vehicle ever lived up to its potential or was understood by residents to be a major concession from the Army. Interest in town meetings and attendance were limited, and, with occasional exceptions, the Town Council did not prove a strong spokesman for resident concerns. Much more influential was a second civilian political component, the Central Facilities Advisory Council created in March 1944. This committee was to share responsibility with the Army for all policy relating to community operation, and, while that viewpoint was not binding on the Corps, it consistently had substantial influence on military decisions. Not unlike the Recreation and Welfare Association, which also influenced the Army in townsite operations, the CFAC was composed of representatives from the several major operating companies on the area. This meant that throughout the war period the primary civilian perspective provided to the Army was not democratic but corporate in nature.

Manhattan District officials were willing, therefore, to make concessions in the interest of resident good will, but there were limits. The likelihood of compromise dwindled sharply when questions revolved around possible espionage or information leaks. Corps efforts in these areas had great effect on the lives of residents and commuters, though neither knew the full extent of that influence. The primary unit for assuring secrecy throughout the Manhattan project was the district's own Intelligence and Security Division created especially for the task. Although independent of Army Intelligence and the Federal Bureau of Investigation, the division was closely supported by them.[15]

The Intelligence and Security Division was headquartered at Oak Ridge, and in addition to uniformed personnel it included approximately 500 agents operating in civilian clothes. Agents were young and hand-picked on the basis of their background. They carried out a variety of assignments, acting as special couriers, as body guards for General Groves and other individuals whose safety was deemed vital to the project, and as ever watchful guardians of information related to MED operations. In the latter capacity, agents were often drawn all over the world in pursuit of rumors and to plug security leaks. A letter from an American officer on the front lines in Belgium to a friend in Oak Ridge innocently inquiring whether the latter was working on atomic energy could, and did, bring an agent to Europe.[16] Another set off to examine the rumored skin rash of an ex-MED employee who was in South America. The ex-employee had voiced a belief that the rash was a consequence of working with a "queer ray" while at Oak Ridge.[17] Even Superman felt the force of the Intelligence and Security Division pressure when his creators began to put commentary into the comic strip about "atom-smashing cyclotrons."[18]

Because of close cooperation between Intelligence and Security agents, general military intelligence, and the FBI (and presumably also because of the secrecy with which each operated), it was seldom possible from Oak Ridge records to determine which agency was at work in a given instance. What was clear, however, was that intelligence activities were quite widespread and extensive. Some of those activities were local counterparts of the broad "security leak" missions of the Intelligence and Security Division just described. In one case this meant rushing agents to the small town of Maryville, some fifty miles from Oak Ridge, to query a local minister who had used the atom as a subject for a Sunday sermon.[19]

Another category of intelligence activities was general supervision of security precautions on the reservation. Illustrative was a decision attributable to the Intelligence and Security Division to install a com-

Some entrances to buildings, such as this one to K-25, were
plastered with admonitions and cautions.

pletely new system of identification badges for reservation personnel in early 1945 after it became clear that a large number of badges passed out under the prior system could not be accounted for.[20] A regular action whose jurisdiction was less clear was to test directly the viability of security arrangements. Agents would deliberately attempt to enter the reservation, or its prohibited areas with improper identification. At times such penetration missions proved shockingly easy. Certainly this was the case with the report of one dismayed operative in April of 1945. He walked freely through the gates in the prohibited K-25 area and subsequently in and out of the restricted buildings, all the while carrying a large camera, flash bulbs, and a tripod. Much to his own horror he was even able to photograph openly classified buildings and was never asked for a pass.[21]

Needless to say, the ultimate intelligence concern was to assure uninterrupted project operations and to identify possible espionage situations. The former effort not only called for vigilant scrutiny of daily operations but cautious anticipation of potential future problems as well. This effort could take such minute form as quietly notifying Tennessee Eastman Company in 1946 to terminate an employee believed to be a sexual deviate. While his case history suggests he was not vicious or likely to cause trouble, the company was told, "it is impossible to predict when sexual perverts of this type will become dangerous." A similar reaction followed the arrest of a cement finisher in early 1945 as a sodomist. The police file reported that Mr. Ford admitted to being a sodomist, "was born that way and is unable to do anything about his unnatural craving." In order to prevent any further disturbances (he had threatened his boyfriend when he found the boyfriend in bed with a girl), the reviewing officer concluded that "he must be eliminated from the area." Ford was escorted through termination proceedings and off the area and barred from being re-hired.[22] The same fear of disruption caused intelligence forces to develop very strong contingency plans when it appeared that the reservation might be hit by a labor strike in 1944. One measure was to build up quickly the number of project guards and firemen who were quartered in barracks on the area. More than 250 such individuals were moved in and quarters were provided free of charge. There they would be readily and quickly available in the event of emergency. A second quiet arrangement was made with the Army's Fourth Service Command. Troops would be dispatched immediately from Camp Forrest, Tennessee, should a labor crisis arise.[23]

Concerns over possible espionage and sabotage at Oak Ridge also caused security forces to involve themselves in such diverse actions as surveying the construction of TVA's Norris Dam to the north of the project and urging purchase of land on the west bank of the Clinch River that bordered the reservation. In the case of the former, the question was whether sabotage at Norris would bring destructive floods to the Clinton Works. Ultimately it was determined that flooding would not reach the main plant areas and that the dam was so massive in construction total destruction was unlikely anyway.[24] In the case of the latter, high ground across the river at one point provided an excellent location to view the progress of construction in the K-25 area. The location was attracting many interested visitors, and there was no assurance they were all friendly locals. Twenty-eight hundred new acres were promptly added to project holdings and the high ground closed to spectators.[25]

Finally, espionage fears encouraged security forces to establish a careful monitoring system over project personnel. Presumably, though available records offer no proof, one dimension of this activity at Oak Ridge was some degree of surveillance by plain-clothes agents of the Intelligence and Security Division. These agents were active throughout the Manhattan District. Another dimension was countless investigations of specific individuals or situations carried out by Roane-Anderson at the request of intelligence forces. Most interesting, however, was an arrangement established by security agents wherein reservation residents and other employees watched each other. The induction procedure was simple and standard. A project employee would be contacted by intelligence and called in for an interview at an office on the reservation. Once there, an intelligence officer would simply remind the selected individual that operations at Oak Ridge were of major importance to the war effort and that tight security was an absolute necessity; the officer would then request the employee's assistance in the security effort. The new operative was pledged to secrecy and asked merely, in the course of their normal daily activities, to be sensitive to suspicious actions, apparent security breaks, or other matters which might be of interest to intelligence forces. Such information was to be included in brief "chatty" letters mailed at specified intervals in previously addressed envelopes provided by intelligence to the offices of the Acme Credit Corporation in Knoxville. It was not possible to establish how many Oak Ridgers acted as intelligence operatives in this way, but such evidence as does exist suggests that number was substantial.[26]

While townsite residents and other employees might be generally unaware of the extent of these intelligence activities, they were acutely aware of security precautions expected of them. Never for a moment did the military allow them to forget it. When new residents entered the reservation, the general information bulletin they received clearly stated that expectation on page one: "You are now a resident of Oak Ridge, situated within a restricted military area. . . . What you do here, What you see here, What you hear here, let it stay here." The demand for secrecy and the fear of espionage was not, of course, unique to Oak Ridge MED during World War II. Workers and residents on every military and defense installation were warned of the

Outposts, strong fences, and armed guards—here at Y-12 and
K-25—were commonplace in security-conscious Oak
Ridge.

151

need to guard against loose talk and the unauthorized disclosure of information to the enemy. The level of both security and concern with MED was, however, probably significantly more intense than in the nation at large.[27] Among the first acts required of new employees in Oak Ridge, as elsewhere, was to sign an official Declaration of Secrecy certifying awareness of the need for secrecy and of the penalties under the Espionage Act for violation of security. Employees who later resigned or were terminated from the project were again reminded of the Declaration in exit interviews. Delivery of townsite resident directories was accompanied by warnings that the document was classified. Admonitions against loose conversation were run repeatedly in the *Oak Ridge Journal*, which stated at the top of each issue "Not to be Taken From the Area." Project personnel were also constantly bombarded with a poster campaign stressing the need for secrecy as well as given periodic security talks at their places of employment.

Almost anything could become an object of secrecy requirements. Seemingly minor matters, such as the desire of some residents to follow the common practice of people away from their permanent home in sending relatives copies of church bulletins from services attended, became an issue of security.[28] A number of items found in many homes in civilian communities became the focus of security concerns and required registration by on-site residents with the military. These items included cameras, field glasses, telescopes, and firearms. In a significant commentary on divisions within the town, the guns were flatly prohibited in the trailer and hutment areas.[29] Even death certificates and official letters of sympathy to relatives of employees accidentally killed on the project became classified documents and were not delivered to next of kin until after the war.[30]

The precise success of the efforts at secrecy among residents and commuter employees is difficult to establish. Two generalizations do seem warranted, however. One is that many more people probably possessed degrees of information on the purposes of the project than was commonly portrayed in postwar reporting on Oak Ridge. Second, secrecy admonitions were respected by reservation personnel to a surprisingly high degree.

In the case of the former, several levels of awareness may be identified. Presumably, though evidence was insufficient to establish it, there were individuals based at Oak Ridge who were aware of full project operations. Most assuredly, however, that number was very small. A far larger group with less, though substantial information on the project was scientific personnel. Perhaps on the thesis that complete and accurate knowledge was less dangerous than speculation within this group, such individuals were routinely and fully informed of operations and goals of the plant at which they worked. As a general rule this group was not informed of activities in other plant areas. In fact, however, at least some awareness of purposes at all three major plants was not uncommon, especially among scientists at the X-10 pile. Beyond this group, access to information was on a need-to-know basis in which the official level of knowledge for each employee was clearly indicated on project badges. Whether or not this provided an enhanced status to those wearing specially numbered badges is impossible to determine from the record. Under this system the great majority of reservation personnel officially did not know the nature of the activity at Oak Ridge, though many had access to certain amounts of sensitive information about aspects of project operation.

In this case of the latter group, that information was apparently adequate for a significant number to guess correctly at least that overall concern was with uranium atoms. At the Oak Ridge Public Library the dictionary page upon which the word "uranium" appeared had to be replaced repeatedly because of finger wear.[31] Var-

Some of the men most responsible for developing the atomic bomb are pictured in 1940; *from left,* Ernest O. Lawrence, Arthur Holly Compton, Vannevar Bush, James B. Conant, Karl T. Compton, and Alfred L. Loomis.

iations on the experience of one project employee
whose duties caused him to travel across Tennessee
were probably not untypical. Repeatedly asked by reservation friends to obtain for them on his next trip to
Nashville a certain volume that dealt with uranium and
atomic energy, the employee finally read it himself.
Because of his own work-related knowledge, he subsequently understood with reasonable accuracy what
the project was about.[32]
   On the other hand, it was more surprising that secrecy was maintained as well as it was. There were
always rumors, and in numbers sufficient to concern
Manhattan officials, but, in general, secrecy worked because project personnel were willing to abide by it.
Those with the most information apparently said the
least, generally refusing to confide even in their
spouses. The closest neighbors or friends often did not
know exactly where or in what capacity the other was
employed. It quickly became a matter of principle as
well as community custom not to ask employment-
related or project-related questions. Reservation personnel apparently became even more closed-
mouthed about project operations when dealing with
people unconnected with Oak Ridge. The fact that
the revelation of the actual nature of the project following the Hiroshima bombing was such a surprise to
residents of areas adjacent to the reservation, as well as
to the great majority of Oak Ridge residents and other
employees, was of course testimony to the success of the
secrecy policy.
   The final chapter in what became the story that linked
forever the histories of Oak Ridge, Tennessee, and
Hiroshima, Japan, began in December 1944. In that
month General Groves officially informed his superior that it appeared "reasonably certain" a bomb would
become operational at some point in 1945. The
estimate was that one would be ready August 1, 1945,
and a second by January 1, 1946.[33] At this point
Groves foresaw no need for preliminary testing. His

optimism seemed warranted. Already in the fall of
1944 the atomic pile at MED's Hanford site had been
activated. In Oak Ridge, Tennessee Eastman had
taken over the first production building of the giant Y-12
electromagnetic plant from the builders in early 1944.
After some initial problems, the K-25 gaseous diffusion
plant become operational in February 1945.

Needless to say MED officials intended to do everything possible to assure that all continued to go well.
One potential threat to that operation, union desires to
organize Oak Ridge workers, had finally been resolved in 1944. Union officials ultimately agreed at a Washington meeting with General Groves and James F.
Byrnes, then administrative assistant to the President,
that organizing efforts would be halted for the duration of the war. In the eyes of the MED, unionization was
a dangerous threat to security. Union leaders, individuals over whom the military would have no direct
control, would be sure to acquire substantial amounts
of classified information in the normal course of their
work. Beyond that, the Army simply could not allow the
kind of free discussion of project activities which was
likely to occur among workers at union meetings.[34]

A second matter of major security concern even by the
fall of 1944 was the rapid progress of the war in
Europe. The fear was that successful conclusion of that
struggle would encourage significant numbers of reservation workers to reduce their efforts or even leave
their jobs completely. As early as September 1944,
hardly three months since the Normandy landing, a
military propaganda campaign to "educate" workers
against such actions was underway. The closer the
apparent collapse of Germany came, the more concerned and nervous Oak Ridge authorities became. The
"education" campaign was gradually intensified in
the reservation newspaper, in billboards and streamers,
and via direct instructional activities among employees by project operating contractors.

When it did come, V.E. day was to be carefully

orchestrated. By pre-arranged plans, news of the German surrender would be conveyed to residents and workers through sound trucks and through military notification to operating contractors who would in turn inform employees. Both presentations would stress the need for continued effort until the defeat of Japan. Already prepared handbills emphasizing the same theme would be rushed to the chief of the Roane-Anderson guards and the chief of the fire fighting units for distribution at pre-established points across the townsite. One concern of the military was simply temporary disruption of operations at CEW as a result of "unrestrained hilarity and possible violence" which V.E. day might engender. Perhaps because of this the Corps encouraged churches to hold religious services on that day. Indeed, designated churches, time of service, location, and attending ministers had been worked out before the end of October 1944. Fears about consequences attendant to that event proved largely unfounded and most employees stayed at work. "Stay-on-the-job" propaganda activities, however, went on at a steady pace through the remainder of the war.

The Army plainly feared that with the defeat of Germany, Oak Ridge workers might lose their patriotic fervor and drift off to jobs that held more postwar promise. The campaign to keep them at work, which showed a rather distinct lack of faith in the popular support for the war effort, had many facets. A film shown in area theaters in April 1945 featured Under Secretary of War Robert Patterson commending Oak Ridgers and urging them to complete the project with all possible speed. The *Oak Ridge Journal*, in announcing the film, reminded readers that the importance of the project would not end with the defeat of Germany. "Remember Pearl Harbor, Corregidor, Bataan, Tarawa. . . . We have begun to repay Japan for the mass murder of helpless civilians and prisoners of war and we will not quit until they are defeated." In July of that

year, the *Journal* warned anyone planning to leave
Oak Ridge that there were heavy layoffs and extensive
unemployment in the Mid-West and the Great Lakes
areas while there was none at CEW. The work there
would continue full force until Japan surrendered.[35]

The official V.E. date was May 8, 1945, Harry Truman's sixty-first birthday. Suddenly thrust into the
presidency by Franklin Roosevelt's death on April 12,
Truman had received his first complete briefing on
operations of the Manhattan Engineer District by
Groves and Secretary of War Henry Stimson only two
weeks after taking office. The new President was an-
guished by what he heard, and he hoped through the
month of May to find viable alternatives to an atomic
strike against Japan. Truman, however, allowed plans
to continue for a test explosion of a plutonium device in
the uninhabited desert near Los Alamos, New
Mexico.

Two days after V.E. day, the job began in Oak Ridge
of putting together official publicity on the Manhattan
project, materials that would be released to the press
presumably after a nuclear bomb had been dropped. By
early July, details had been carefully worked out for
the handling and distribution of that story. Lieutenant
George Robinson, formerly of the *Memphis Commercial Appeal*, established on the reservation a restricted
reproduction room where a dozen members of the
Woman's Army Corps detachment prepared mimeographed information sheets on the bomb. By the end
of the month, thousands of mimeographed pages of fourteen separate press releases had been reproduced,
and Intelligence and Security Division agents had been
dispatched with releases to key Southern cities to
await the actual dropping of the bomb. Releases for the
Knoxville media were held in readiness at Oak Ridge
itself.[36]

On July 16 in a stretch of desert fifty miles from
Alamogordo, New Mexico, the efforts at Hanford and
Los Alamos reached their planned conclusion in a

flash of atomic fire. Until then Manhattan scientists had simply never been sure their plutonium device would actually explode. Success of the so-called "Trinity test" was immediately conveyed to President Truman, then attending the Potsdam Conference with Winston Churchill and Joseph Stalin. Slightly over a week later and still away from Washington, Truman tentatively approved the first atomic strike against Japan. It would occur as soon as weather would permit a visual bombing run after August 3, 1945, on one of four targets: Hiroshima, Kokura, Niigata, or Nagasaki.[37]

Final decisions were made against a background of growing concern among scientists associated with the Manhattan project about the social, political, and moral implications of this use of atomic power. Through the early summer some scientists sought to establish a series of formal meetings on the subject. MED's Arthur Holly Compton, a top scientist on the project, tried to convince General Groves that the meetings could have positive consequences. Groves did not agree: discussions posed potential security problems and they would be halted.[38] A less formal gathering did take place on June 28 at Oak Ridge, however, and subsequently the names of those who attended were reported to military authorities in Washington.[39] A second meeting occurred at X-10 in the wake of news leaked back about results of the Trinity test. It was occasioned, however, by X-10 officials seeking to persuade scientists not to sign a petition sent down from MED's Chicago site. It urged that the destructive force of the bomb be demonstrated to the Japanese but not actually dropped on inhabited areas.[40]

On July 25, the last shipment of Uranium 235 necessary to what would become the Hiroshima bomb left Oak Ridge, arriving at its destination, the Pacific island of Tinian, on July 27. Indeed, the so-called "Little Boy" bomb would use all the U-235 then available. While plants at Oak Ridge continued to operate on a twenty-four hour schedule, it would take weeks of additional time before sufficient material for a second uranium bomb could be ready. The Navy's USS *Indianapolis*, carrying most of Little Boy's component parts, reached Tinian on July 29. In a curious twist of history the *Indianapolis* was sunk four days later en route to the Philippines by a Japanese submarine.[41] Already on Tinian was the crew of the B-29 bomber, "Enola Gay," who had been chosen to carry out the atomic mission. They did not know, however, until informed by the flight commander, Colonel Paul Tibbetts, Jr., on the evening before the attack that the force of the explosion would probably equal 20,000 tons of TNT.[42]

Things became almost shockingly simple at the end. As if by a giant act of metamorphosis, over three years of work involving a peak employment of 82,000 people and a cost of $1,106,393,000 in Oak Ridge alone was suddenly transformed into the activities of a single bomber crew. The "Enola Gay" departed Tinian at approximately 3:00 A.M., Monday, August 6, 1945, Tinian time. Fearing the possibility of a crash on take-off, final bomb assembly was only completed once the flight was underway. Hiroshima had been designated as the primary target with alternate targets at Kokura and Nagasaki should weather force a change. It would take the strike ship and her two observer escort planes about six and one-half hours to cover the 1,700 miles to the primary target.[43]

The three B-29's arrived over Hiroshima slightly after 9:00 A.M. Tinian time or approximately 6:15 P.M. August 5, Washington time. There was a clear straight attack run of about four miles. The drop was made from 31,600 feet. At 9:15 a crewman pressed a switch and released the single missile. Less than sixty seconds later there was a flash which became a purple fireball; two shock waves hit the plane and a seething mass of cloud rolled up over 40,000 feet to form a mushroom. The escort bombers snapped limited photographs as all three fled, only losing sight of the massive atomic

From Y-12, operated by Tennessee Eastman, came much of the U-235 that made the "Little Boy Bomb" possible. *Inset:* Shown here is an interior shot of the Y-12 beta vacuum system.

cloud 363 miles from the target. Hiroshima was devastated. Approximately 1.7 square miles extending out one mile from ground zero were essentially cremated. Over 96 percent of the buildings in the city as a whole were destroyed or heavily damaged. Estimate of casualties by the Japanese was 71,000 dead or missing and 68,000 injured. At 9:20 A.M. Tibbetts had radioed Tinian: "Mission successful." In retrospect it rings a hauntingly ironic response.[44]

In Washington an anxious General Leslie Groves received word of the event about 11:30 P.M. Truman, still en route from Potsdam, was quickly notified and efforts got underway to issue an official announcement from the President. He had radioed approval from the USS Augusta. Approximately 10:00 A.M., August 6, Eben Ayres, assistant press secretary, informed reporters in the White House press room that "There might be something later." At 11:00 A.M., newsmen were met by Major General Alexander D. Surles, War Department public relations officer who had the official announcement. Ayres read the first paragraph, then handed out copies: "Sixteen hours ago an American airplane dropped one bomb on Hiroshima. That bomb had more than 20,000 tons of TNT. It is an atomic bomb."[45]

Three days after the first strike, a second drop was carried out at the city of Nagasaki. This time the weapon was "Fat Man," the plutonium bomb developed from materials at Hanford. It took 35,000 lives. If the Hiroshima attack could be justified as a grim but necessary demonstration of power intended to force an immediate end to the long bloody war, the Nagasaki bombing was far less defensible morally. Whatever the actual effect of the event, however, the desired result followed on its heels a few days later. A stunned Japanese Empire surrendered on August 15.

As William Manchester has pointed out so well, much that seemed insensitive and even vulgar in the response of Americans to the use of atomic force must be put down to incomprehension. The concept was simply too big. It could only be absorbed gradually.[46] When that did begin to happen there would be debate about the decision to drop the bomb and there would be substantial concern about future implications of this new destructive capability. More typical of the time, however, were the large number of letters which began to reach MED officials in Oak Ridge in the days immediately after Hiroshima.

There was the incredible naïveté of an Arkansas resident who wished to purchase some atomic bombs "of the right size" to blow out some stumps in his field. His letter was addressed to the "Atomic Bomb Company" at Oak Ridge.[47] There was the almost endless effort by commercial organizations to associate themselves with the work of the project. Plywood manufacturers wished to know what role that product had played in construction of community homes.[48] A laundry equipment concern asked permission to advertise that its equipment had been used in the townsite laundry.[49] A chair producer asked to apprise the public that its products had been placed in reservation quarters.[50] A MED officer wrote a reply to an Atlanta, Georgia, company cautioning it not to convey the impression in advertising that Merita bread was the only bread distributed at Oak Ridge.[51] Even Yale University wanted in on the act. New Haven newspapers had implied that the University played no role in the development of the atomic bomb. A representative of the chemistry department wrote to obtain clearance of a press release which stated that MED research had taken place in Yale's Sterling Chemical Lab from January 1942 to August 1945. The announcement would be made by the president of that institution.[52]

The reaction of Oak Ridgers was similar to that of the public at large. They had been told that the project was vital to the war effort, and they were pressed constantly to complete it, even though most did not know its real nature. Suddenly the secret was out.

They knew what they had produced; it had proven successful in operation; and they were informed that it was instrumental in the conclusion of the war. For the people of Oak Ridge, August 6 was the day they would remember. News of Nagasaki and indeed to some extent even the Japanese surrender were anticlimactic.

Because the news of Hiroshima was broadcast, those persons with access to radios heard first. Ironically, this meant that even many scientists and employees who helped to produce U-235 received the news when wives telephoned to tell them. That could be a shock. "I felt," recalled one employee, "that when my wife said some mention had been made of the atom, she might be splitting up herself."[53] Initially, conversation was hushed and limited. Neighbors gathered at homes in small groups awaiting further details. Employees were still reluctant to say much to each other about the news. Sworn to secrecy for so long, it took a bit of time to comprehend the change that had occurred.[54]

By mid-afternoon, however, whistles, horns, and any other devices that could make noise were ringing out across the reservation. People left their homes and jobs and began to gather in the central business district of the community, where almost continuous celebration went on for the remainder of the day. "Extras" of Knoxville newspapers sold for $1.00 a copy. One circulation man sold 1,600 papers in thirty-five minutes.[55] There were exceptions. A wife rushed to the townsite hospital to confront her recuperating scientist husband: how could he take part or involve his family in a project that would bring the death of innocent civilians, she demanded to know.[56] Another wife felt disappointment; she had become convinced that the "energy" which was being produced at Oak Ridge was in fact powering American bombers on their missions. Such a new energy source could be a major and permanent contribution to a peacetime civilian

Above: Oak Ridgers celebrate V-J Day with understandable enthusiasm. *Below:* Colonel Kenneth D. Nichols received a Distinguished Service Award for his work as District Engineer for MED.

world. A bomb was merely a weapon and of much less importance.[57] Some residents, suddenly frightened by the presumed dangers of living in a location where bombs were produced, actually began to pack up and leave Oak Ridge.[58] Yet all such cases remained minor exceptions to the general mood of hilarity and self-congratulation.

The August 6 celebration could also be understood as a significant rite of passage, though most residents probably did not understand it that way. In theory, at least, the reason for the very existence of Oak Ridge was gone. The demands of top secrecy which had so molded the nature of daily existence in the community had changed fundamentally. The project and the town had suddenly become a focal point of great public interest whereas only hours earlier the preference, at least of the military, was that no one should know either existed.

Perhaps the magnitude of the change was best expressed by an apparently minor occurrence on the morning of August 7. As the Southern Railroad's train from Washington rolled into the Knoxville station, a Pullman porter announced to his passengers: "You are now entering Knoxville, the gateway to Oak Ridge."[59] The future might be unclear, but it was going to be very different from the past.

6. Oak Ridge – Atomic Capital of the World

On September 6, 1945, employees of CEW were officially assured through a previously prepared news release from Colonel Kenneth D. Nichols, the top ranking Corps officer at Oak Ridge, that the existence of the reservation was in no immediate danger. A postwar program for the operation of the project was being placed into effect, and the program would "insure continuation of most reservation activities," pending a decision by Congress on some permanent plan for control and development of nuclear energy. Some limited changes and cutbacks in certain operations, of course, would be necessary in the upcoming months, Nichols admitted.[1]

The announcement was an important one. For one thing, it seemed to be the first official recognition by the Army that Oak Ridge or the reservation as a whole had any future beyond "the duration," though exactly when the military actually put aside their original understanding of CEW as merely a temporary wartime project cannot be established in the written record. A second significant aspect of the announcement was its recognition that their employees were understandably uneasy about what might occur at the reservation. Corps efforts to combat and minimize these fears would become a major concern the year following the Japanese surrender. Residents would come to recognize, however, that while military assurances were sincere and honest, they were not always totally accurate. The "limited changes and cutbacks" noted by Nichols proved substantial.

These "changes" were not the only matters which would cause significant uneasiness among Oak Ridgers and employees. Another was the action, or more properly the inaction, of the United States Congress on the matter of nuclear development. Future administration and the long-term implications of atomic energy became a legislative concern in the fall of 1945. While understandably a complex question, deliberations dragged on in what too often appeared to be an uncertain, halting, and indecisive fashion for almost a year before the Atomic Energy Act became law in August 1946. In the meantime continuous and often ominous rumors about the future of Oak Ridge swept through the reservation almost daily. The federal statute passed in August did not end those rumors. It placed control and development of nuclear power in the hands of a newly-created Atomic Energy Commission, which would assume the responsibilities and obligations held by the Manhattan Engineer District, effective January 1, 1947. By this action Oak Ridge was to pass into the hands of civilian authority, but precisely what that meant for the future remained most unclear.

If this uncertainty were not enough, residents watched while many of their neighbors packed up and departed from the reservation. Some left because they became too apprehensive about the future of CEW to remain. Others went simply because they had never seriously considered staying beyond the duration of the war. Finally, many left because they were laid off in postwar employment reductions. Whatever the reason for departure, it surely must have appeared to those who did remain that CEW and the town were dissolving before their eyes.

In May 1945, the reservation had reached its peak employment of approximately 82,000. The residential community was also at its peak, approximately 75,000. Within three months after the Hiroshima bombing, employment had fallen to 51,000, and the community population was down to 52,000. By January 1946, the figures were 43,000 and 48,000, respectively. In June 1946, employment dipped to 34,000. The community stood at 43,000. As might be expected, heaviest employment cutbacks were among construction workers where by June 1946 no more than 2,000 of a once peak force of 47,000 remained on the project.

At least some of the women who found jobs at Oak Ridge during the war would be under special pressures if they tried to keep them. In the summer of 1946 Captain William Bonnet, deputy chief of the Facilities and Service Division, asked the chief of the Civilian Personnel Branch how he could go about removing the female employees of the motor pool. During the emergency, he said, the majority of those hired had been women. Now that the emergency was over, he thought it highly desirable "and in the best interest of more efficient operation of the motor pool," if all females could be replaced with males. He claimed that it had been difficult to require female employees to perform first-echelon driver maintenance.

The period had also taken its toll on operating personnel as well: 10,000 left between May 1945 and June 1946. Nor did the problem of population instability end with the summer. By September, employment was 2,000 under what it had been in June. In the first three months of 1947 Tennessee Eastman alone laid off 5,000 employees. The 1950 Census would report the population of Oak Ridge at 30,205, though this number was still sufficient to protect the town's previous ranking as the fifth largest city in Tennessee.[2]

For all these problems, there were certain positive aspects about life on the reservation, at least by the summer of 1946. A new and pronounced community spirit was observable in the town. One sign was that auto plates bearing the inscription "Oak Ridge—Atomic

Above: Grove Center, shown here in 1947, continues to be a major shopping area for residents in the west end. *Below:* Jefferson Square, to the west of Grove Center, is another retail center in Oak Ridge.

Capital of the World" had become common fare on the cars of community residents.[3] At first glance an occurrence of no consequence, in fact this was a happening of substantial significance. It suggested that many residents had begun to think of themselves as "Oak Ridgers" rather than merely "making do" as temporary inhabitants of an impersonal federal project. That was new. It suggested also that for all their apprehensions, many residents had come to believe that the town did have a future in postwar America.

Why this apparently new commitment to Oak Ridge should begin to emerge in a period of such equally apparent disorder is difficult to say. Ironically, it may be that the seemingly disruptive departure of so many people actually increased, in subtle ways, the life satisfactions of those who determined to remain, thereby increasing their loyalty to the town. As more than one observer noted, once out from under the heavy population and production pressures of the war, the community began to look much more like a normal town.[4] Indeed, released from these pressures the Army and the Roane-Anderson Company in late 1945 began to turn their attention in a concerted way to landscaping, repainting, and cleaning up the town. In general, as the journal *Business Week* took note, Oak Ridge was losing its previous rag-tag frontier appearance.[5]

The very decision made by residents to stay in Oak Ridge, of course, provided a form of community with their neighbors not characteristic of the town in the past. During the war period, "home" was always elsewhere. At the same time the decision of many wartime dwellers to depart brought an increased social stability which many residents may have found attractive. The nature of the exodus was to reduce considerably the disproportionate number of single adults, especially the more rootless, less educated, construction workers. By the end of 1946, 47 percent of the Oak Ridge population was made up of families. In these population changes, however, the community did continue to retain its previous youthful atmosphere which had been attractive to many dwellers. Even in 1948 only 19 percent of the town's residents were over age forty as compared to 35 percent for the nation as a whole.[6]

Finally, it may well be that community spirit or sense of satisfaction was increased because residents as a whole were living physically in better accommodations than ever before. By the summer of 1946, approximately 80 percent of them were located in houses or apartments. This matter of housing was a significant one, for it symbolized graphically the changes in the type of concerns the Army and community faced once the war was concluded. Throughout the war period the issue of housing was clearly the most difficult problem in the community. The limited amount available on the project combined with the complexity of the allocation procedure to promote anger, charges of favoritism, and persistent controversy. The problem was in fact never totally eliminated in the postwar MED period. By the summer of 1946 there were still 2,300 eligible applicants for single family houses on the reservation.[7]

The decreasing number of residents and thus those who desired to be quartered on the reservation did, however, reduce the extent and "heat" of the housing issue significantly. Dorms were a case in point. Whereas in the war period two and three people had been assigned to single rooms, private rooms once more became available. (Parenthetically, either in anticipation of the reduction in pressures on dorm residents which these changes would make or because it had not obtained the anticipated result, the program of dorm counselors instituted in 1944 was ordered terminated as early as August 1945.) Eight dorms were actually closed out in early October 1945. Also that month the Army moved to eliminate the white hutment area. Inhabitants were assured of "much better living quarters" in dorm space. Over the next several months the number of trailers in service as living quarters was drastically reduced, the majority of them apparently being returned to the Federal Public Housing Authority from whom they had been borrowed.[8]

Even in the case of the difficult to acquire single and multi-family houses where availability never met demand, regulations over who might receive them were significantly eased. Enlisted men with families became eligible and a greater number of such units was allocated to officers of the Corps. Allocations to concessionaires were increased. Previous rules that excluded housing for individuals who lived within forty miles of the reservation were also relaxed, as were minimum salary requirements requisite to application for on-site houses. Clearly too, the Army was increasingly willing to make minor exceptions to existing housing regulations.

Moreover, with general housing pressures on the project reduced, even reservation blacks came in for a belated share of attention. Throughout the war they had been confined to hutment dwellings with a few of these accommodations ultimately remodeled and allocated to families. In October 1945, Roane-Anderson proposed to the Central Facilities Advisory Council that 350 housing units be made available to black families. Applicants would be thoroughly screened to assure "that the colored community would be a credit to Oak Ridge." The aging dream of the Negro Village had never been completely forgotten. After some discussion the council recommended that Roane-Anderson should initially set aside approximately 150 prefabricated homes on Scarboro and Raccoon roads, as well as an additional 47 units on Arrowwood Road.[9]

The Scarboro-Raccoon area was in the vicinity of the existing black hutments, while the Arrowwood location was toward the eastern end of the reservation near Elza Gate and the originally proposed location for the Oak Ridge black community. Single blacks, male or females, continued to receive quarters only in the hutment areas.

Whites then living in the newly designated black locations would be moved into "other suitable housing." It would appear that adequate homes were readily available to accommodate displaced whites. By the end of 1946, additional commitments of prefab houses had been made for black families. At that point well over 200 units had been blocked out for "the colored." In an unusual variation on normal procedures for housing assignments, it was officially agreed in January 1946 that applications for homes by Negroes would be screened through the Colored Camp Council, and houses would be assigned directly by Corps officials rather than by individual operating companies.[10]

While concessions to black Oak Ridgers were hardly adequate to the need, two distinctly new patterns were clear in the Army's relations with reservation blacks following the close of the war. One was the greater advisory role given to the Colored Camp Council. This group came to be consulted carefully on virtually every matter, however small, affecting reservation blacks. The second was a much greater concern for the quality of the life of the blacks and a willingness to improve it (though in retrospect efforts seem meager, they did represent significant change). The liberalized policy on houses was a case in point. Other examples would be assignment in March 1946 of a surplus hutment bathhouse to the black community, assistance in renovating that structure into a facility suitable for church services, an increase in the number of movies provided each week for blacks, and inauguration of a "Colored Area Nursery School" in still another vacated hutment bathhouse.

The military's increased responsiveness to the concerns of Oak Ridge blacks did not grow solely out of reduced housing and space pressures. After all, the Corps had long shown an interest in developing a black community on the reservation. Perhaps the close of the war really did provide the first practical opportunity to take some solid initiative in this area. Perhaps also significant was the fact that reservation Negroes were more aggressive and better organized than they had been for much of the war period, and therefore their desires carried more force than previously. It may be as well that the times made some difference. With the days of MED now limited, it could be that the Army felt its own self-protection demanded more careful consideration of conditions of life among black Oak Ridgers since President Roosevelt's 1941 Executive Order 8802 forbade racial discrimination on all defense projects. Moreover, concern about civil rights was now much in the air as witnessed by Truman's establishment of a National Commission on Civil Rights in 1946. Finally, perhaps the change was partly a matter of publicity. Concern with the conditions of life among blacks on the project had prompted a visit in December 1945 by representatives of the powerful black newspaper, the *Chicago Defender*, and in the initial rush of news coverage of the project following Hiroshima, CEW had usually obtained a certain amount of "bad press" because of the deplorable conditions in the hutment areas.[11] Yet in the end, it seems very probable that surplus space, much more than ideology, was at the heart of the Army's postwar posture toward Oak Ridge blacks.

As noted, the problem of adequate housing on the reservation was never solved during the postwar MED period but certainly was reduced in magnitude from its previous proportions. For this the Army surely must have sighed with relief. On the other hand, what MED officially lost with the conclusion of the war was the crisis-motivated posture of most Americans to accept deprivations willingly and to moderate their personal ambitions for the duration—or at least to give the appearance that they had done so. In many ways it was a poor swap, as Oak Ridge authorities would soon learn.

In the immediate area of CEW, hostility toward the reservation increased, if anything, over previous levels. After all, the reservation was supposed to be a war

175

project. But the war was over and Oak Ridge was still there. That it was likely to remain was not a generally happy realization for many locals. In 1947 it meant that Anderson County legalized liquor sales, although an overwhelming number of county residents outside Oak Ridge were opposed to it. This was only the beginning, as many long-time residents surely recognized. Criticism which had largely abated during the war among residents of Oliver Springs, Clinton, and Knoxville about denial of access to Highway 61, a road which crossed the Manhattan project, again became heated. As one Oliver Springs resident put it squarely to his congressman, "apparently some of the Big Brass of CEW don't know the war is over."[12] Presumably with that struggle concluded, the government would not need much of the land it had taken in Roane and Anderson counties. When would previous owners have the opportunity to buy it back, many asked. "Probably never," even when couched in the best bureaucratic jargon, was not a response happily received.

Nor were matters much better in terms of "willingness to sacrifice" inside the reservation. Charges of inequities flowed in vast numbers to authorities. For example, one resident bitterly resented his failure to obtain a new furnace unit simply because those available went to "favored 'key personnel' hand-picked by authorities." There had always been some abuse of concessionaire rental specifications. Corps officials finally determined, in January 1946, that erroneous reporting of gross receipts by commercial concerns to reduce their monthly rental payments had become so widespead that the Pinkerton National Detective Agency had to be hired to investigate and expose the practices.[13] Nevertheless, a resident demanded redress for the "injustice" she suffered in non-renewal of a concession agreement in 1946 to operate a book and gift store; terms of her contract be damned, she contended, termination was more appropriate for someone else.

Two apparent threats to the efficient operation of CEW were more serious from the point of view of reservation authorities. Both threats were vastly overestimated. One was a degree of dissension that had grown throughout the American scientific community about the future use of atomic energy. In Tennessee this degree of dissension was minor; some reservation scientific personnel joined the Association of Oak Ridge Engineers and Scientists, a group concerned about the ethical implications of their work.[14] These individuals had no desire to disrupt operations at Oak Ridge, but they did wish, along with other dissenting scientists, to educate the public about the destructive potential of atomic energy and to promote plans for the international control of nuclear power. Such dissension, except for limited debate about MED purposes following the New Mexico test blast in 1945, was both new and threatening to the military.

The second "apparent" potential threat to operations at Oak Ridge arose in the rigorous and competing efforts by the American Federation of Labor and the Congress of Industrial Organizations to organize project workers. This issue would replace housing problems as the No. 1 concern of the Corps in the postwar period. The problem of unionization was not, of course, a new one; efforts by the AFL unions to organize workers had begun as early as 1944. Indeed, in that year, two groups, the International Brotherhood of Fireman and Oilers and the International Brotherhood of Electrical Workers filed petitions with the National Labor Relations Board in Atlanta to be recognized as official bargaining agents on the project. The NLRB proceeded to schedule public hearings on the claim of the unions, but at this point the Army intervened. MED had many reservations about this labor activity. The Army feared that, at best, organization efforts could jeopardize project security and, at worst, might result in a strike which could delay the successful completion of the atomic mission. More specifically

When the war was over, the drive to unionize became
the No. 1 concern of the Corps. Shown here is an AFL
organizer at a 1946 rally.

with respect to the pending petitions for unionization, the Army indicated great concern about the public discussion of the project which that process would involve. Faced with an appeal to wartime loyalty and warnings about endangering the crucial nature of CEW operations, union officials withdrew their petitions "for the duration."[15]

When the war was over in the summer of 1945, the drive to unionize was once more in full force. The Army, having already anticipated the move, was not surprised. At least as early as January 1945, the MED Intelligence and Security section had worked out general guidelines for reservation unions, and the guidelines would remain essentially the same for the remainder of the Manhattan project period. Essentially there were six principles: (1) Labor groups would be permitted to hold meetings in Oak Ridge under the same conditions that were imposed on other groups in order to guard the security of military information; (2) all union members and local officers must be employed by CEW contractors; (3) only CEW employees would be permitted to attend meetings on the project (this was later modified on a per person basis to allow entrance to various speakers and agents of union national offices); (4) all meetings on the project would be open to an officer representing the district engineer (theoretically at least, this was to assure that classified information was not discussed); (5) union files, financial accounts, and other records would be open at all times to inspection by the Army; and (6) all mail and other communications between local CEW unions and outside agencies or persons, including the office of the National or International Union with which they were affiliated, would be subject to censorship by the Corps. Finally, in July 1946, the review of union literature and correspondence before distribution was terminated, though certain prohibitions on the types of information which might be provided to outsiders were continued.[16]

Union organizers and members were hardly pleased with these requirements. More than once, members grumbled openly in meetings about the presence of a security officer. The members would doubtless have been more unhappy had they realized that even union meetings in Knoxville, which appeared to touch on affairs in Oak Ridge, were regularly monitored by "confidential informants" of the military, as well as by agents of the Army Counter Intelligence Corps. Whatever the discontent of union representatives, however, they received few concessions. Consistent with policy long applied to all groups in Oak Ridge, the union organizers were not allowed to solicit door-to-door, to use loud speakers, or to employ other "public nuisances" such as hillbilly bands. There would be no parades, at least near the residential areas, because of the traffic hazards.

Over and above these restrictions, the unions had further problems. The Atomic Trades and Labor Council tried to purchase an ad in the *Oak Ridge Journal* and was informed by both the paper and the Labor Relations Department of the Corps of Engineers that no advertising space would be sold to any labor organization. Complaining that these prohibitions did not apply to concessionaires, churches, lodges, the Red Cross, and the American Legion, the AFL protested to the district engineer, enclosing in their letter a check for $6.00 and the ad copy. The request was denied.[17]

Beyond these general provisions on the reservation, unions would be sharply curtailed in the ways they might distribute promotional literature. Ultimately the Army would insist that even literature distributed at the project gates must be handed out by bona fide employees of the project and that they must do so while wearing their identification badges. Although there is no evidence to substantiate the charge, union leaders regarded this practice as a way of identifying members of the companies which employed them. Workers would then, they feared, be subject to retaliatory action.

Labor groups would also operate under strict procedures in picketing, should that occur. Pickets outside
the gates would not be allowed to block approach
roads and would be subject to constant surveillance for
violations of Tennessee law, which would then be
reported to appropriate local authorities. A picket who
was not an employee but who attempted to enter the
project with a visitor's pass would be ejected promptly
from the reservation. The Army would also provide
protection to project employees, if they chose to cross
picket lines.

An especially interesting aspect of Army administration in these regulations, however, was a strong
determination to avoid a physical confrontation between the military and union members off the reservation. For example, in illicit distribution of literature,
use of loud speakers, or participation in unapproved
parades, Oak Ridge police were informed by the
Corps that, if demonstrators refused to desist, "no
further action should be taken." Even in the matter of
escorting employees through picket lines, troops should
make no attempt to break up the lines. In both cases,
Army authorities would make an official protest to the
appropriate union. Outside the reservation, picketing
problems would be left to county authorities.

The military was much less generous when it came
to union disturbances or attempts to picket on the reservation itself. Corps orders had been somewhat vague
on these points, implying that some peaceful on-site
picketing might be acceptable. In August 1946, however, General Groves came upon the policy statement
that the "military prefers no picketing or strikers on
the area." He responded quickly and hotly that "there
will be no picketing in Oak Ridge," he stated, "and all
concerned should be governed accordingly."[18] As that
order passed down the line from the top military
authority at CEW, it was translated to mean that if picketing did begin and was not halted in twelve hours, it
would be stopped "using all forces at your disposal."[19]

There were those in the Oak Ridge officialdom who questioned the constitutionality of such a flat prohibition on "peaceful picketing" as a possible infringement on the right of free speech. Grove's order stood, however.

On the other hand, the precise and legally acceptable reaction of the Army to a strike and picketing on the reservation was of continued concern to Manhattan officials. The matter was never really resolved. The heart of the problem was the unclear jurisdictional status of the reservation between the federal government and the state of Tennessee. Apparently all CEW authorities agreed that project officials could eject strikers and picketers, using such force as was reasonably necessary. Beyond this, however, there was grave doubt about the legality of using military forces to quell any civil disorder that might occur. These were normal police responsibilities that traditionally belonged to state and local governments. Commitment of troops would be warranted only if these authorities could not or would not provide adequate protection. In the end the military determind that troops would not be used against strikers unless the threat to life and federal property was of such an imminent and emergency nature "so as to render it dangerous to await instructions from the War Department."[20] The agreed upon approach to civil disorder on the project, whether strike motivated or not, would be to call upon state and local authorities for assistance. In the meantime, using the right of "citizen's arrest" provided for in the Tennessee Code, the Army could commit to service the immediately available portion of the auxiliary military police and regular military police employed on the reservation. If necessary, and this contingency was provided for, troops could reach Oak Ridge from Tennessee military facilities in a matter of hours.

Actually such concerns proved quite unnecessary. Indeed, the more serious struggle was not between CEW plants and the unions but between the AFL and the

CIO. By summer 1946, six national unions—three AFL, two CIO, and one independent—were seeking NLRB recognition. The military was hardly enthusiastic about unionization or the inter-union struggle at Oak Ridge, but by March 1946, it was at least fully resigned to both. In that month the NLRB was officially notified through the War Department that it was now possible and consistent was national security to develop procedures jointly, under which the board might handle labor union cases at the Clinton Engineer Works.

Both the AFL and the CIO considered the struggle for power on the project one of major significance. The Clinton reservation stood as a primary battleground in the larger issue of which federation would ultimately become the spokesman for Southern labor. Competitive efforts were intense and often bitter. Heated charges and countercharges over unfair, if not illegal, practices among the several involved unions were almost constant. The CIO claimed repeatedly that the Army favored the AFL. While there is no evidence that military authorities actually overtly acted in ways to favor the AFL, it may well be that the sympathies of local authorities were with them. Army security officers, not unlike certain other federal agencies, were suspicious about possible communist influence in the CIO hierarchy. Beyond this, it was AFL unions which had so readily made concessions on organizational activity at what the Army considered a crucial moment in 1944. Finally, as MED officials were quite happy to acknowledge and as the secretary of war publicly stated in a 1946 Labor Day speech, the AFL's International Brotherhood of Electrical Workers Union had worked diligently and had been enormously helpful during the war in recruiting craftsmen for Oak Ridge, Hanford, and Los Alamos.[21]

Yet if, in the end, the organizing efforts of the AFL at Oak Ridge proved more successful than those of the CIO's locals, it had far less to do with the attitude of the Army than the greater effectiveness of the drive by the

AFL on the workers at CEW. Records of the struggle between the two labor federations make it clear that AFL unions worked harder, longer, and in a more organized fashion than did their CIO opponents. It should be noted too, however, that the AFL began their postwar efforts with a significant advantage over their CIO competition. Many members of the AFL came to the project as construction workers, and then stayed on as plant employees once the town and the other installations had been completed.

On the other hand, in the summer of 1946, the results of the initial vote and subsequent run-off for unionization could hardly be considered a significant victory for either federation. In the first election, AFL unions obtained a higher vote than their CIO opposition at all three major plants: Tennessee Eastman (Y-12), Union Carbide (K-25), and Monsanto Chemical Company (X-10). The largest group of eligible voters (44 percent), however, either failed to vote or voted against representation by unions of either federation. In the subsequent run-off, employees at TEC voted against union representation. At Union Carbide, the CIO edged out the AFL by twenty-five votes, though the latter did keep a solid majority at the smaller Monsanto operation. Approximately 17 percent of the eligible voters failed to vote. It was hardly a decisive verdict. Both federations vowed after the elections to redouble their organizing efforts at the project.[22]

The Army was at best unenthusiastic about union activity on the reservation and, in retrospect, overly concerned about the dangers of it. Yet for all the precautions and regulations, it is significant that the military did accept without apparent opposition a wide range of union operations in 1946. The Corps boasted that in the postwar period it had greatly enlarged the civil liberties of reservation residents, and, despite such actions as the clandestine monitoring of certain off-site union meetings, there was a clear and pronounced new tolerance within the Corps administration of Oak Ridge in the postwar period. In this regard it is quite noteworthy that MED officials on the reservation processed routinely in 1946 the application of certain project employees to organize a chapter of the Socialist party, USA.[23]

Yet in one sense the new tolerance did not represent any great change in military desires for life at Oak Ridge. Normalization had been a wartime idea, and it remained so in peacetime. In this respect the main difference was merely that the degree and speed of the effort could now be greater. As early as January 1946, reservation officials seriously considered removing the gates, thus allowing free entry and exit to the community, but this was not accomplished until 1949. The officials were also actively studying, as the year progressed, what type of political identification or organization might be the most appropriate form as the move toward greater normalization proceeded. The two primary options in the eyes of the military were that of a federal district analogous to Washington, D.C., or a fully incorporated municipality under the laws of Tennessee. Interestingly enough, a significant problem in approving the latter option was that Tennessee law made municipal incorporation contingent upon the request for it by at least one hundred legal residents who owned land. Since there were no private landowners in Oak Ridge and no plans to provide for such an innovation, "city" status appeared unworkable. Army officials assigned to make the study concluded, for this and other reasons, that the more preferable status would be that of a federal district.[24]

Even as early as mid-1945 several major community functions previously carried out by the giant Roane-Anderson Company were turned over to private firms as part of the normalization process. It had always been the plan, stated Corps officials, that Roane-Anderson would directly operate certain services only until experienced firms in these areas could be identified and contracted.[25] In fact, these efforts were in major part designed to increase the amount of private enterprise on the reservation, a direction that would be continued by the Atomic Energy Commission after the Army had departed.[26] In the winter of 1945, American Industrial Transit, a new firm organized by executives of the Southeastern Greyhound Lines, took over transportation operations of the huge CEW Bus Authority. With the summer, additional major changes took place. Tri-State Homes began management and cleaning of on-site housing. Cleaning of other buildings was assigned to the Community Service Company. Leatherman and Alley, a company which had managed dormitories for TVA at Fontana Dam, took over supervision of twelve dorms in Oak Ridge. Galbreath and Moore Company would soon assume supervision of additional dorms once the war ended. Gibson Management Service undertook management of the Gamble Valley, Midtown, Ridgeway, and Hilltop (the last two formerly the Ford, Bacon and Davis and J.A. Jones) trailer camps.

Beyond these changes in housing and transportation, Roane-Anderson would also be relieved of a variety of other responsibilities in the summer and early fall of 1945. A private concern was contracted to distribute ice on the reservation, ninety tons every day in the summer and half that amount in winter. Distribution of coal and fuel oil to the trailers and hutments went over to civilian hands. So too did the collection of garbage and trash. By October 1, Canteen Food Service, a Chicago firm, had taken over management of the white and black hutment cafeterias as well as the white hutment canteen. A Virginia firm would soon sign a contract to operate the project laundry. Even pest control and bottled water services were turned over to private concerns.

The changes continued in 1946. The Army decided that the drab and largely uniform exteriors of townsite houses would be painted in colors selected by tenants. Roane-Anderson received continued encouragement

from the military to contract for commercial concessionaires with an eye to "improving services through
competition."[27] The giant Oak Ridge phone system,
previously run directly by the Army, was placed in the
hands of Southern Bell. Even the reservation police,
already formally separated from the Guard Force in
March 1945, were further reorganized to make their
civil and criminal responsibilities essentially those of
any metropolitan force. It was more than coincidence
that their uniforms were changed from military khaki
to the conventional municipal blue.

Oak Ridge's vaunted medical system went through
its own "normalization" process. The previous Corps
effort to maintain a doctor-patient ratio of one physician per 2,000 population was dropped. Planning for full
conversion from military to civilian medical practice
was under way as early as September 1945. Actual transition began in February 1946, when dental service
was put on a civilian basis. A second step occurred in
March when the reservation's civilian psychiatrists
departed and it was determined that the psychiatric
unit would be terminated once the military personnel
assigned to it were discharged. Also the Oak Ridge
hospital would be converted to private or municipal
ownership. The Public Health Department would be
separated out from the hospital, however, and set up
with its own administration as a public function similar to what was the case in other cities.

As medical officers were discharged, they were temporarily replaced with civilian doctors who were employed by Roane-Anderson. This was a transition state
which was to last only until the military could be
assured that there would be a level of private practice
adequate to the needs of Oak Ridge. On March 1,
1946, civilian physicians were allowed to begin private
practice on the reservation and, with the exception of
continued management of the hospital, which would
pass into private hands several years later, Roane-
Anderson withdrew from involvement in the health care

of residents. In September 1946, the transition to civilian medicine was further underscored symbolically when the Army authorized a name change from the "Hospital Clinic Building" to the "Medical Arts Building" for the structure that housed the offices of many of the new doctors. Even the old, highly comprehensive, prepaid medical insurance plan, originally developed because security prohibited disclosure of information required by private companies to write a group policy, was terminated. The Corps decision to disengage itself from the current plan forced the Oak Ridge Health Association to seek coverage from a private carrier. They ultimately selected the Tennessee-based Provident Life and Accident Insurance Company.

Although the Army's accelerated normalization efforts brought significant changes in life at Oak Ridge, the small town model by which the military had always defined and sought to develop the community remained constant. Thus the Army continued to regulate closely all matters which would affect the attractiveness of the town's physical environment. This concern would mean a myriad of widely varying decisions by the Army, from specifications for chicken coops and their acceptable locations, to careful supervision of signs used by reservation retail stores so as to avoid the growth of "garish" commercial advertising.[28] Concerted efforts were made to minimize the bureaucratic aspects of life on a federal reservation. The Army continued to maintain a low profile and measures which acted to personalize the relationship to municipal agents of government were encouraged. An excellent illustration of the latter was a practice introduced in May 1946, whereby residents leaving the project for vacations or other long trips were invited to notify the Oak Ridge police of their impending absence. A patrol car would then check these houses on a regular basis and, if desired, even mail residents a weekly postcard informing them of the condition of their homes.[29] As of old, the military remained highly supportive toward social clubs even to the extent of making exceptions on the previous general prohibitions against releasing membership lists to an organization's national headquarters.[30] Clearly the type of club most favored was the civic service organization, the mainstay of small town life. An MED official made the point clearly on the occasion of the inauguration of a rotary club in 1946: "To me nothing is more indicative that normalcy is being rapidly attained here than the presentation of the charter here tonight to the newly formed Oak Ridge Rotary Club, for Rotary activity is a part of our normal American way of life."[31]

By the fall of 1946 some apprehension among Oak Ridgers about the future remained, but it was much less than it had been during the previous twelve months. They now knew more about the details of the civilian administrative take-over slated for January 1, 1947. Although they could not be sure of the ultimate implications for the city of control by the Atomic Energy Commission, the Army at any rate continued to assure residents and employees that there was no reason to believe that drastic change would be forthcoming. The rumor mill certainly did not cease to operate, but it did appear that Corps officials and those of the AEC were working closely together to assure an orderly and easy transition.

Indeed, as the fall progressed Oak Ridgers seemed increasingly almost blasé about the change, if the absence of information in the town weekly newspaper is any index of the popular mood. The impending transition was seldom mentioned in the *Oak Ridge Journal* except for special occasions such as the November visit of the five Atomic Energy Commissioners to the reservation. Even here coverage was limited and in straight news form.[32] Only one brief note on the upcoming event appeared in December, the transition having been edged out by matters of more immediate interest—Christmas events and local news. The January

2, 1947, issue of the *Oak Ridge Journal* carried a one-column front page article titled "CEW Transferred to Atomic Energy Commission Rule," which was largely comprised of comments from Major General Groves and David Lilienthal, the AEC chairman. Page 4 also contained a very brief editorial which congratulated the Army for a job well done and extended best wishes for success to the AEC.[33]

Twenty miles away in Knoxville news coverage and apparent popular interest in the transition were even more scant. Very rarely through the fall of 1946 was the event or even Oak Ridge mentioned in either the *Knoxville Journal* or *News-Sentinel*. Here too in the month of December, Christmas and local news dominated coverage. While December 7 might seem a likely point for some reference to Oak Ridge, no coverage or comment occurred. The front page of the *News-Sentinel* was dominated by news of a fire in Atlanta's Winecoff Hotel, which took the lives of 114 people. Even limited comments given over to Pearl Harbor were relegated to page 3. The January 1 issue of that paper did contain a page-3 news article on the AEC transition at Oak Ridge, noting that most military personnel there would remain at their assignments until suitable civilian replacements could be found.[34] The *Knoxville Journal* on that day gave the event abbreviated comment on the front page, though it was far overshadowed by much greater coverage on the twelve new "beauty queens," one for each month of 1947, recently chosen by the city's Esquire Club.[35]

So with minimum notice in the Oak Ridge area, the AEC quietly obtained control from Manhattan authorities of a collective land area which exceeded in size the state of Rhode Island. A few months later the Manhattan Engineer District ceased to exist. It all seemed so simple. Quite promptly on January 1, 1947, the five-year Army-civilian experiment in community development and control at Oak Ridge came to an end. And, indeed, the town would ultimately be the better for that loss. Yet reflecting on the Corps of Engineers' all pervasive impact in the community during these years and the permanent influence of this period on the history of the city to date, there is surely a high degree of poignancy in one seemingly minor occurrence during December 1946 in the life of Leslie Groves. He had received a query about job possibilities at Oak Ridge for a civilian who wished to move there. Taking note of the area's imminent transition to the Atomic Energy Commission, the remainder of Grove's reply was largely routine elaboration on that point. It concluded with apparently unrecognized irony: "Accordingly, I am no longer in a position to be of assistance to anyone seeking employment."[36]

Epilogue:
Oak Ridge Revisited

On a warm September day in 1942 five men stood near the railroad whistle-stop of Elza in Anderson County and looked west over the long valley before them. Two were high ranking officers in the Army's Corps of Engineers. The other three were officials of Stone and Webster Engineering Company. It was noon. In a little while the five would enter a car and make the twenty-mile drive through rural countryside to Knoxville. Once back at the Andrew Johnson Hotel one of them would call Washington, D.C., "The location tentatively selected in July and which has been under survey is ideal for our purposes," he would report to superiors. "We should proceed with land acquisition."[1] It was upon this recommendation, backed up by a personal visit to the site, that Leslie Groves made his decision to locate a nuclear project in East Tennessee.

The history of the Clinton Engineer Works and the Oak Ridge Community in a significant sense, therefore, began at Elza in 1942. That same area is also a fitting point to conclude the present volume. At least it seems so to the two University of Tennessee historians who on a spring day in 1979 parked briefly at the approximate location of the old Elza entrance to the federal reservation. That gate once sat astride Highway 61 (later the Oak Ridge Turnpike), which ran west directly down the valley that became Oak Ridge toward the K-25 plant area at the opposite end of the reservation. The two historians were reflecting on their newly-completed manuscript that examined the history of the community from 1942 to 1947. What the manuscript lacked were appropriate concluding remarks. In the end what could be said about the significance of the town's wartime birth, and the relationship of that birth to the city's present and future? The two had come to the old Elza gate to look afresh at Oak Ridge, to ponder, and, they hoped, to find those words.

MEMORY ONE
From the Dispossessed

An elderly man at the time of the interview, his family had come to Anderson County about 1919. His father wanted river-bottom farmland and purchased approximately 600 acres adjacent to the Clinch River. The land was important. It meant security for the family. As a young adult he continued to work the farm but also opened a country store. The MED took both. He first became aware of the government's intent to do so via an article in the *Knoxville News-Sentinel*. The price received for his holdings was unfair, he believed. There was particular irony in this forced transaction. His father had sold off some of the family's land at atypical depression prices some time earlier. Because this was virtually the only sale in the area during recent years, Army appraisers used it as a primary precedent upon which to set condemnation land values in the acreage absorbed by the MED. When officially notified, the family was told to vacate in fifteen days. They were given no reason for their sudden expulsion. "It wasn't right," he recalled with a degree of emotion.

Like most of the dispossessed, he and his family tried to relocate in the nearby area, in this case, Clinton. There was no assistance from the MED. In 1944 he received a concessionaire contract from Roane-Anderson to operate a gasoline station inside the reservation near the Elza Gate. Unable to gain on-site housing he commuted daily from

Clinton. Not unlike most residents of Clinton, he was not
happy with the continued postwar existence of Oak Ridge.
He had hoped the land might be returned. He was
sympathetic with the feeling in the county that Oak Ridgers
were all "a bunch of Yankees," or at any rate "outsiders." As
with many local people, he saw no reason why they should
have the right to vote in county elections. They were not, after
all, really a part of the county and they paid no local taxes.

In retrospect, he held no real malice toward the Manhattan
project or Oak Ridge. Actually, he was proud of the city's
wartime nuclear role and his relationship with it. When
reminded of the destruction at Hiroshima during a visit to
Japan some years later, he simply replied in non-defensive
fashion that Americans also still remembered the attack at
Pearl Harbor. He himself had long ago returned to the Oak
Ridge area to live and over the years had fared well. In
general he believed that the changes forced on Anderson
County by the existence of Oak Ridge had been positive
ones. But the way they took the land wasn't right, he repeated
periodically. That verdict was both his past and present.[2]

It was sunny and warm as the historians began their
journey along the Oak Ridge Turnpike from the Elza
Gate entrance. Ironically, the first business
establishment on the left-hand side of the road was a
Yamaha motorcycle dealership with its home offices
in Japan. Directly across the street was a liquor store.
Each in its own way clearly symbolized the distance
of contemporary Oak Ridge from the war experience.
Behind the Yamaha dealership was the World War II
vintage Louisville and Nashville rail spur that ran parallel to Warehouse Road southwest toward the Y-12
area. Storage buildings from that era intermingled along
the road with more recently constructed warehouse
and small commercial facilities. Just beyond the liquor
store on the right was a large sign board that proudly
boasted the emblems of fifteen civic clubs including the
Atomic City Toastmasters. Oak Ridgers were still
joiners.

Moving down the turnpike from Arkansas to
California avenues, the historians noted that the old

alphabet system of street names remained intact. Indeed, it was still being employed, they had observed
earlier, in at least some of the very recent affluent
subdivisions, such as Emory Valley which was several
miles to the southwest near the Clinch River. While
trees and other obstacles obscured the view, to the right
in what once consitiuted East Village and the area
behind it up the ridge to East Drive was a body of homes
which looked very much like remodeled small
cemestos. Appearance was deceiving, however. The
original and very hastily built East Village and adjacent housing were made up, with limited exceptions, of
prefabricated TVA-designed flattops. They had once
housed over 1,500 families.[3]

Understood to be of lesser quality than cemestos at
the time of their construction and built at densities far
greater than desirable, these dwellings in general
were deteriorating badly by 1952. Indeed, much of this
housing had been either torn down by the Atomic
Energy Commission or sold for removal. In that year, a
program was begun to redevelop the larger East Village area with single-family homes of far better quality
than the flattops, a program which was also designed
to reduce the previous structural density by about one-
third. A limited number of flattops remained in this
area, as well as in the western portion of the city, until
the summer of 1957 when the last sixty-six such
dwellings were removed.[4]

The original cemesto houses actually began farther
down the turnpike at California Avenue, though extending into the East Village area via the intersecting
Alabama Road. The least well kept of these units were
the multi-family dwellings still located on their World
War II sites at periodic intervals, one and two blocks off
the turnpike on the right and running parallel to it.
Single family units covered the side of the Black Oak
Ridge north to Outer Drive and west in varying density for slightly over two miles. These homes remained
generally well kept and this portion of the community

has continued to be a highly desirable location in which to live. Most owners, however, had long ago covered the cemesto surfaces of their houses with brick, some type of wood, or aluminum siding. Many had also added carports or garages, extra rooms, and patios.

MEMORY TWO
From the Ridge

He had first come to Oak Ridge from Lexington, Kentucky, in July 1943 to join the electrical division of Tennessee Eastman Corporation. She followed a month later when a two-bedroom cemesto home became available just off Tennessee Avenue. For her the move was not a happy one. She expected little except a variety of hardships and privations which must be suffered until the war's end. Yet for all its restrictions and peculiarities they both immediately fell in love with the town. While others barely endured such things as the interminable mud, they found constant delight in strolls along the boardwalks to and from Jackson Square. They enrolled their two children in kindergarten and first grade (a third child would be born later) and prepared to enjoy the "duration" on the best terms they could make.

Nor were those terms bad ones at that. Ultimately both held full-time jobs and worked hard, but for them life in the townsite was always exciting and interesting. If they had few very close friends, they had many acquaintances and led an active social life. He noted with regret at the time of their interview that the wartime custom of spontaneous unannounced visits among friends had gone by the board. Neither recalled any sense of social distinction based on the type housing one occupied (many living outside the cemesto area would disagree). More socially conscious than he, she did recall with resentment that Army officials were always able to obtain the best housing—"it was common knowledge." Both agreed that possession of a phone had substantially enhanced their status among those who knew of it.

They were never a part of the group who were anxious to leave Oak Ridge. It always seemed, she mused, that the people most vocal about their desire to depart the reservation,

when questioned, had the least to go back to. Whatever else
they felt at the bombing of Hiroshima, it sharply threatened
their personal world. He feared the project might be
dissolved and they would be forced to move on. She hoped
that somehow they would be able to stay. They did indeed
stay, see their children grown, and in his case, retire after a
long, rewarding, and productive professional career.

Talking with them in their present remodeled and attractive
cemesto home, it was clear that if there had been bad times
in the Manhattan period, the two had long forgotten them.
Now there was only nostalgia and warm affection in their
recollections. It was a good time and obviously still very much
part of their lives.[5]

The first stop the historians made was at the reservation's original Guest House. That structure was still in
place and appeared to be in good condition. It was
now the Alexander Motor Inn, but its wood-frame construction and structural similarity to so many other
remaining war-period dorm buildings betrayed its original purpose. The building was not unattractive even
now, with its large porch dotted with rocking chairs and
running the length of its front. In the war years guests
standing on that porch could look out over tennis courts
just down the hill on to what surely must have seemed
a sea of dormitories. Although virtually all of the dorms
were gone, the tennis courts remain. If guests in those
days looked slightly to the left across the Oak Ridge
Turnpike, they could see the barracks-like reservation Administration Building. "The Castle," it was
called. The latter has now been torn down and replaced by a modern federal office building housing the
Department of Energy. Back behind that structure,
however, there remains a series of war-period buildings, which presently serve largely as records-storage
areas and offices. From the Guest House the historians
moved to the Chapel on the Hill located, as of old, just
behind that structure. It also appeared to have fared well
over the years. A small, white frame structure with a
spire characteristic of traditional church architecture in

pre-war America, the chapel sat on a grassy knoll surrounded on two sides by cemesto homes, and the chapel obviously had been carefully cared for. It still housed the United Church, an interdenominational congregation formed during the war. A third landmark of the area—the community's first high school once located across the street from the chapel—had fared less well. While aware that a new high school had been constructed farther along the turnpike, the historians noted with some shock that the old school building had been completely razed. A driveway to nowhere and a parking lot now remain as anachronistic monuments to an almost empty site—the football stadium on the original site, however, is still in use.

The old Guest House and the Chapel on the Hill were located just off Jackson Square, which was originally designed to be the single major commercial area for the reservation. That vision was, of course, soon destroyed by rapidly accelerating demands for more and more project personnel. The historians would return to Jackson Square a bit later in the day. For the moment, however, their destination was Outer Drive, which ran along the top of the ridge. In a real sense their trip west on that artery provided a fundamental recapitulation of the community's history. The large and spaciously laid-out homes of the more eastern, and earlier constructed, portions of Outer Drive began to give way to much smaller homes and yards. Beyond New York Avenue these small cemestos in turn gave way to even more closely spaced, frame prefabs, most of which were simply rectangles in design—many appeared to measure no more than 20' by 20'. The quality of original construction was obviously poor, and there was little concern with decorative location of houses on lots or with landscape architecture in general. They represented a concluding chapter in the story of the townsite's westward development as the Army sought desperately to meet the mounting population pressure.

Old Oak Ridge ended abruptly at Louisiana Av-

enue. Once across that street the historians were suddenly very much in the present, having entered one
of the several expensive, attractive, and highly mani-
cured suburbs which had come to characterize the
west end over the past two decades. To the south, where
Louisiana intersected with the Oak Ridge Turnpike,
the line between old and new was less sharp. In the war
period, both west and east from mid-way down this
avenue to the turnpike only duplexes, dorms, and
apartments were visible. Some of these structures
still remain but many had been replaced with single-
family units on Louisiana, and on the turnpike they
were interspersed with an ever growing number of
commercial establishments. Only a short drive west
along the turnpike, however, the affluent new suburbs
would again make their appearance. Arriving at the
Louisiana junction with that highway, the historians
chose to turn back to the east, heading for Jackson
Square. Approximately a mile and a half from that inter-
section they again crossed Illinois Avenue. Had the
year been 1945, the two recalled, they would be entering an area that included on both sides of the road
large trailer camps—"Midtown Trailer Camps," they
were collectively called. A bit farther up the turnpike
and on the right would have been the white hutment
area. The black hutments were located approximately
a mile to the south of the white huts, along what was
then Scarboro Road. Both types of housing had long
since disappeared in favor of single-family homes and
commercial establishments. Among the latter was a
large modern shopping facility known as the Downtown
Center. There was a certain irony in the name, the
historians agreed. Unlike most cities, Oak Ridge had no
traditional "downtown," only the long turnpike and
the string of stores that clustered along or within a street
or two of it. The pattern was a logical consequence of
the town's original design—yet another MED legacy.

The drive up the turnpike was a study in incongruity. The Colonel's Kentucky Fried Chicken, Burger

King, and other fast-food establishments nestled in unsettling mixture amid a variety of MED-period dorms now converted to office and business space. Housed in one such structure were the purchasing offices of Union Carbide's Nuclear Division. Shortly before the necessary left-hand turn to reach Jackson Square, a roadside sign boldly pointed in the direction of the graphite reactor and informed tourists or other interested parties that it was a mere ten miles away at the Oak Ridge National Laboratory. For the two historians steeped in the "secrecy" ethos of the MED period it was almost shocking. X-10 had become an advertised historical landmark. Driving north on Tyrone Avenue, they also noted with surprise that two of the MED-period dorms, Canton and Charleston halls, were still in use as housing. "Lodging for Men, by the Day, Week, Month" the sign read.

MEMORY THREE
From the Dormitories

Six community women held a luncheon reunion at the Mayflower Restaurant. In the war years they had been residents of Beacon Hall, a dormitory which along with Beaumont and Bayonne had recently been scheduled to be torn down. That decision was the occasion of their gathering. Today, any recollections of hardships that life in the dorms imposed seemed absent from their memories. If they had suffered loneliness or depression (certainly common maladies in the dormitories), if they had resented the crowded living conditions of that setting or the many restrictions on their personal life, such feelings were certainly not evident on this occasion.

They had each come to the reservation young and largely inexperienced and were taken with the excitement of the place. Assignment to a dorm had not troubled them. Actually, it was a very good place to start friendships. This was important, and part of the charm of Oak Ridge. Everyone was away from home, as one noted, and friends were like a surrogate family. Beacon residents enjoyed each other; the laundry room in particular became a kind of social hall for them. The ratio of men to women was twenty to one, another fondly reminisced. So there were plenty of dates. Although few cars were available, it mattered little—there were always the reservation recreation facilities, including dances at Ridge Hall and at the tennis courts. Sometimes, the six recalled, couples simply rode the reservation busses and kept getting transfer tickets so they could sit and talk in privacy. Very near Beacon was Boone Hall, a men's dorm. That had morale advantages too. Five of the six had ultimately married Boone Hall men.[6]

In a more private setting and at a moment which perhaps lent itself to more serious reflection, a seventh community woman also recalled MED dorm life. She too had come to the project young and inexperienced, in this case from Kentucky. For a month she was housed in a single barracks-like room at the Guest House with twenty other women. She slept on a cot and there was no place to store her possessions. It was always noisy, drinking was heavy, and theft was epidemic. She was unhappy and almost daily on the verge of returning home. Finally a dorm room became available. Another woman in the Guest House asked to be her roommate. She refused. The prospect of her own space and privacy had become too precious.

The dorm was better, though her room was very small, and, if there was less theft than in the Guest House, it was still a common occurrence. She, too, remembered the laundry room, though less for the social activity there than for the time spent almost nightly in long lines to reach the large sinks to wash the mud, inside and outside, from one's boots. They tried to provide entertainment on the reservation, she recalled, but there was very little to do. Besides, the area was not well lighted and she at least, was a bit frightened about getting out and doing too much wandering around at night. She went home to Kentucky every weekend and felt fortunate that "home" was near enough to do so. But she made some very good friends in the dormitory. They would get together almost nightly, cook their own dinner in one of the rooms (against regulations), play bridge, or find some other diversion—perhaps just sit and talk. What she and three of those friends came to want dearly was to get out of the dorm and into an apartment. Ultimately this happened; probably because of the flirtatious presentation of their case by one of their number to a housing-office official—"she was so pretty and she really could roll those black eyes." Like most of her friends, resident number seven was a school teacher and given the apparently temporary nature of the project, she felt some periodic unease about how long her position would last. (In fact, however, she would remain with the Oak Ridge School System until her retirement many years later.)

Her memories, at least at the time of the interview, were not quite the buoyant unreserved ones of the other six, but there were common grounds among those recollections. "Beacon Hall was a big happy family," she agreed. Even with some misgivings about dorm life, number seven's final judgement was positive, "I remember it as fun."

Thousands of individuals lived for various periods of time in the dorms during the Manhattan years. Most departed sooner or later, especially once the war ended. But there were many who stayed on, those for whom the MED experience would set the pattern for much of the rest of their adult years. If their recollections do not totally square with the reality of the written record on dorm life, it is nonetheless the "reality" they carry with them and the one through which they filter and evaluate their lives in Oak Ridge.[7]

Jackson Square had not changed appreciably since the war period, and its wood-framed shops did not seem too much the worse for wear. Indeed, looking anew at the way in which the stores were closely clustered with spacious parking areas adjacent to them, the square seemed very much a prototype of later postwar shopping centers. Probably few of the original stores continued to operate, but at least one—the old Ridge Theater—was there and still open for business. The Center Theater once also on the square had been refurbished and had become the community playhouse. At one end of the group of stores was Big Ed's Pizza. That location had originally housed one of the reservation's first drug stores. A large lofty structure, which obviously remained quite similar to its war-period construction design, Big Ed's had become a community institution in

Oak Ridgers saw the beginning of a new world for them
as a small explosion symbolically opened the gates in
1949.

Oak Ridge. It had only been on the Square since 1970, but for a town whose total past is merely thirty-five years, such rapid institutionalization was understandable.

Over pizza and beer at Big Ed's the conversation of the two historians continued to center on Oak Ridge. When did it become that "normal" and "typical" American community that MED authorities had sought to emulate and encourage? Not yet, the two agreed. True enough, the fences and gates that enclosed the city were gone. They came down in 1949, it should be noted, with something less than total enthusiasm by residents. A town meeting especially called to discuss the matter voted against opening Oak Ridge to the general public by a margin of four to one.[8] Turner Construction Company, the parent concern of Roane-Anderson, terminated all contractual responsibilities in the operation of the town on January 3, 1952. In the 1950s also, land in the community was subdivided into individual lots for purposes of private purchase or lease. Oak Ridge became an incorporated municipality under the laws of Tennessee in 1959, and in the next twelve months the AEC gradually withdrew from administration of the community in favor of a full local government by residents. June 4, 1960, marked the town's first "Independence Day."

Yet "normal" or "typical" it was not. For one thing the imprint of the Army remained too strong. The earlier drive around the city assured the two historians of that as a physical fact. But there was far more to the matter than merely the physical aspect. Oak Ridgers have continued to define the identity of their town and themselves in terms of the wartime mission. Many of their basic social institutions were also molded in that period: church congregations, clubs, leisure activities (e.g., the community theater group and the symphony), and schools. The city school system remained among the best in Tennessee, and this excellence was directly traceable to the original efforts of the MED.

Moreover, the Army on its wartime mission set in place another pattern that the city has never transcended. In a very significant sense Oak Ridge remained a "company town." The plants at Y-12 and K-25 continued to survive and provide extensive employment based on nuclear contracts, some still involving classified activities. Beyond this, residents clearly saw the future of the city in terms of nuclear research. Seldom did an issue of the community's newspaper, the *Oak Ridger*, fail to mention the topic in some respect. President Jimmy Carter's apparent determination to halt or severely modify the development of breeder-reactor technology and more particularly to terminate work on the nearby Clinch River reactor was not only a severe economic blow but a severe psychological one to the community.

It is, without doubt, partly economic self-interest that leads Oak Ridgers to lead and to adopt quickly what some feel to be dangerous nuclear "progress." But beyond question, the dominant tone of the city remains set by the social and sometimes the political leadership from "the Lab"—X-10, K-25, and Y-12. Working daily with forces considered mysterious and arcane by most outsiders, the scientists sense answers, both immediate and potential, to problems of radioactive waste disposal, possible theft of nuclear materials, radioactive contamination, and other fears voiced by opponents of nuclear energy. In a city that boasted at one time at least of more PhD's per capita than other city in the country, the "high technology"—the highly trained scientific personnel—look to their skills for solutions and trust themselves and their friends to find them.

Coming together during World War II in an attempt to solve heretofore unheard-of scientific and engineering problems and doing it quickly and successfully provided a tradition that seems to have taken deep root in Oak Ridgers. The possibilities, as well as the dangers, inherent in the much-maligned "military-industrial-scientific-labor-academic complex" are, and have been, graphically illustrated in Oak Ridge. With money; a clear sense of crisis and common purpose; military power; industrial, scientific, labor, and scholarly cooperation, the Oak Ridge experience demonstrated its vast and terrible potential.

The MED experience had also done much to fashion the present unique posture of Oak Ridge toward its neighbors. Residents possessed both a very strong sense of community and a significant degree of social myopia when they gazed out toward the limits of the city. They have seldom developed a great deal of interest in the population around them. Since many of those who came to the reservation in the war period were quite different people from the local residents who surrounded the project, they had little in common and knew it. The new residents sensed with some degree of correctness that "locals" looked upon them with suspicion, if not active ill-will. In turn, the view of many Oak Ridgers toward the Tennessee "hillbillies" who inhabited the adjacent county was also far from positive. For residents of Oak Ridge, however, the negative feelings between themselves and their outside neighbors were in the end not very important. Their lives were largely oriented to the reservation and their economic, social, and cultural needs were basically met there. The commonality of the war-time experience plus their own sense of participation in a unique group setting encouraged a strong sense of unity.

In all of these respects the MED legacy has lingered. In a very real sense, one reason why Oak Ridge has not become a typical community is because residents have refused to define it as such. In general they have continued to understand the town as somehow a very different and exceptional place in which to live, much better than their nearby and much larger Knoxville neighbor. While politically a part of Anderson County, they have also continued to have only a minimal interest in the affairs of the county and virtually no

sense of identification with it. The mutual suspicions
and even hostility that once openly characterized the
relationship between county residents and Oak
Ridge residents have become moderated somewhat, although they have not disappeared. Even the modera-
tion has been a recent event.

MEMORY FOUR
From the County

A graduate of Vanderbilt University, he moved to Clinton
from Nashville in 1933. It was just a typical small East
Tennessee town, but there were signs that the city was a
favorable location for a new business venture. Even in this
depression year the local economy was relatively stable. It
also appeared that there would be positive economic
consequences in the area from the activities of the recently
created Tennessee Valley Authority. Although undoubtedly
he did not anticipate it at the time, Clinton and Anderson
County would claim much of his attention and energy for the
rest of his working life. He would in turn become one of the
county's most respected leaders and perceptive observers on
the social and political life of the area.

The creation of the federal reservation to the west came as a
surprise. He like others in Clinton began to hear rumors
about something called the Kingston Demolition Range. The
name sounded ominous. Then suddenly thousands of acres
were being taken to create a "Clinton Engineer Works" and a
substantial number of Anderson County residents were
being displaced by that project without explanation. He felt
the families dispossessed were treated badly. The prices
they were paid for their holdings were unfair. The Army made
no effort to assist in the moves or in finding suitable areas
for relocation. Along with several other local leaders he even
went to federal officials located in Nashville to seek
assistance for those forced to move. It was a curious visit. No
one he contacted was even aware of, or at least willing to
admit awareness of, anything called the Manhattan Engineer
District.

If he did not know the mission of the project in Anderson
County, he certainly knew what it meant for Clinton. The

town was inundated with "strangers" who could not obtain
on-site homes at CEW. They were desperate and would take
virtually any type space. Six men rented rooms in his own
home. A vacated garage in the center of town was
partitioned off into small stalls which were then leased out.
Suddenly gambling, heavy drinking, and prostitution
became social problems. So did disposal of sewage and
provision for adequate schools.

The people of Clinton, he recalled, knew CEW was a war
project and wanted to be supportive, but it became
increasingly difficult. Despite the new governmental services
they required, there seemed no way to tax the "strangers"
for them. Some even wanted to vote in city and county
elections. The laws restricted the vote to those who
intended to make the county their home. Oak Ridgers readily
admitted their determination to depart once the war was
concluded. The people of Clinton were also forced to
compete with the "strangers" for rationed consumer goods
on unfair terms. While local residents abided by the rules, he
believed, not unlike many others in the county, that those
employed at the federal reservation were provided with
virtually as many ration stamps as they wished. "They got
everything they wanted," he recalled with a note of
resentment.

Once in place, the hostility of local people toward Oak
Ridge continued into the post war period. Things were
much better now, he thought. The county was more accepting
of Oak Ridge, and Oak Ridgers increasingly did feel
themselves a part of Anderson. "When they get their tax bill,"
he added with a wry grin, "they know they are part of the
county." There have also been some very good men and
women in Oak Ridge who have provided the leadership for
major improvements in county life, he added.

But there were a number of area people who still resented, at least, the political power of that city in local
elections. "They [Oak Ridgers] can pretty much determine
who the county officials will be." [It was an observation which
obviously was not a totally happy recognition even for
him.] There have also continued to be issues where the line of
division is Oak Ridge versus the county. A proposed
coal-barge terminal on a nearby river was a case in point. To
the county that meant jobs. Opponents in Oak Ridge said,

"it won't be pretty; it will hurt the view; and it will be noisy." We want to protect the environment too, he continued, but we also have to live, and that means jobs [the interviewer noted the significant use of "we"].

Still and all, friction between the county and Oak Ridge was far less now than it had been in the past. That city was no longer the problem it once seemed. He wasn't sure exactly when the change began to occur. Oak Ridge was kind of like a headache, he mused. You know it's there, and you don't stop to think just when it went away, but it's gone.[9]

Whatever the timing, it was clear that in the recent detente between Anderson County and Oak Ridge, the county had compromised or acquiesced a great deal more than had the city. Oak Ridge's sense of specialty and its municipal myopia included a high degree of indifference toward the aspirations of the county. It included also a clear recognition that compromise was not generally necessary. While now less than half the size it was in the MED period, the city dominated the county economically and politically. Because of the volume of economic activity as well as the economic opportunity available there, even in the years of decline immediately after the war, it was inevitable that the commercial life of the county would reorient itself from Clinton to Oak Ridge. It had done so. Politically, Oak Ridgers could, if reasonably unified, carry the day on almost any given local election issue. Of Anderson County's total population which was approximately 60,000 in 1970, more than 28,000 lived in Oak Ridge. The county seat of Clinton had only 4,794 inhabitants.

It is noteworthy, however, that the lines between the city and county populations more recently had become far less clear and sharp than once was the case. This matter had doubtless been one factor in the more recent accommodation between the two. At least in the past decade it had become acceptable, if not fashionable, for affluent Oak Ridgers to buy land and establish homes in the county. While few who had done so might be totally conscious of it, or even willing to admit it, the moves did imply a degree of acculturation to county values, or at least a toleration of them, not characteristic of earlier years. On the other hand, the composition of the city population had also changed over recent years as previous residents of surrounding rural areas found work and moved into Oak Ridge. The scientists and professionals of the city no longer had the nearly complete control they once did. Recent charges that the school system was too oriented toward the college-bound student to the detriment of vocational education were illustrative. That the matter had been able to surface as a viable issue was testimony to significant population changes in the city. Yet whatever were the mitigating effects of these population shifts or the nature of the apparently increasing accommodation between the city and the county, the line between "we" and "they" remained clear. It was likely to be so for the foreseeable future. Still lingering over their late lunch, the two historians could agree upon that, as they had earlier agreed that Oak Ridge had yet to become what might, in any real sense, be called a "typical" town.

Back in the car, the historians proceeded west along the turnpike, turning southeast on Illinois Avenue with Knoxville as their destination. The visit was essentially over, though they did allow themselves two final brief digressions, visits to the neighborhoods of Scarboro and Woodland, both located just off Illinois. Neither was an MED project, but both related directly to significant aspects of that experience. The so-called "Master Plan" for the future development of Oak Ridge, prepared by Skidmore, Owings and Merrill for the Atomic Energy Commission in 1948, projected a city of thirteen neighborhoods. Each was to be based on a desired maximum size in its supporting elementary school of 500 pupils. Only five of the thirteen subcommunities existed at the time of the plan.[10]

Scarboro and Woodland were the first new residential areas constructed in the postwar period. In substantial part, each was developed with the aim of alleviating immediate housing shortages which would indeed be increased in the late 1940s by Oak Ridgers displaced because of an AEC program to eliminate existing substandard housing units in the city. Scarboro was the Atomic Energy Commission's version of the aged MED plan for a Negro Village. Accepted local legend stated that the city's black leaders were called together in 1948 and essentially asked to choose between two areas then being considered as possible locations for a racially segregated black residential community. One was the site originally planned by the Army, East Village. The second was the one-time location of the giant Gamble Valley Trailer Camp. The blacks consulted had chosen the latter.[11] While there was no evidence to dispute the story of a free choice, it did seem likely that the presence of black housing in areas adjacent to Gamble Valley, along then Scarboro Road, Raccoon Road, and the vicinity of Gamble Valley Warehouse Road probably influenced black leaders and white officials as well. The latter at least may also have been influenced by the fact that Gamble Valley was open for immediate construction. The trailers were gone, but sewer lines into that location remained in place, as did a previously constructed elementary school.

In any event, it was anticipated that all Oak Ridge blacks would reside in Scarboro, which indeed they did until the late fifties and early sixties. Scarboro was begun in 1948 and first opened to residents in 1950, at which point 15 cinder-block single-family units, 143 frame duplexes, and 7 dormitories became available for occupancy.[12] A right turn off Illinois Avenue and a short trip down Tuskegee Drive (originally Gamble Valley Warehouse Road) took the two historians into the Scarboro area. The geographic boundaries of that neighborhood remained essentially the same as the old Gamble Valley Camp. Houses, identifiable as constructed under the auspices of the AEC, were small,

modest in appearance, and spaced closely together. As
the historians were aware, problems of substandard
living conditions over the years had been greater in
Scarboro than in other neighborhood locations in the
city. Still, it seemed apparent on the day of their visit that
the area remained alive, well, and a viable community unit.

Back on Illinois and up that road perhaps three-
quarters of a mile, the two turned left, in their second
digression, entering the postwar residential area of
Woodland. The original black hutments as well as the
limited number of family hutments and victory cottages that were built adjacent to them (collectively
known for a brief time in the postwar period as Scarboro Village) were once located in the southwest portion of the subdivision. No signs remained of that
grim aspect of town life.

MEMORY FIVE
From the Hutments

A tall, muscular and imposing black man at the time of
interview, he appeared much younger than his declared
eighty years. He had come to the project from Maryville,
Tennessee, about fifty miles away. There was a strike at the
aluminum company where he was employed. "I wasn't for
it," he recalled. "I just wanted to work." He heard that jobs
were available at CEW and that the pay was good. Both
proved true. Ironically, what also proved true, at least in his
memory, was that racial segregation during the working
day on the reservation was more pronounced and
comprehensive than was the case on his previous job in
Maryville. Moreover, he stated, in matter-of-fact fashion, as
elsewhere in the South the Negro had only two
employment choices—janitor or laborer. He chose the latter
and was promptly put to work pouring cement in the Y-12
area.

For a time he commuted daily by bus but finally decided
to move to the project hutments. His wife remained in

Maryville because on or around CEW, as he put it,
there "just wasn't no 'fittin' place to stay." The reservation was
for him an alien and "mean" setting. Life in the hutments was
at best unsavory. Theft was commonplace. Violence was even
more so. [He seemed to take some small consolation in
recalling that blacks believed the white hutment area—the
"Wildcat" they called it—was at least as rough and violent.]
There was only one cafeteria at which blacks could eat, so if
one missed a meal serving period, "you were out of luck."
He was especially troubled by the reservation security
restrictions. "You had a badge with your picture on it," he
recalled. "You had to show it to come in, go out, ride the
busses, do anything."

Given the violence-prone environment of the hutments and
the racial prejudice of many white southerners working at
CEW, it was easy for a black man to "get in trouble" on the
reservation. He did not "get in trouble." "When I got
married," he recalled, "I told my wife I was going to take care
of you." That became his primary commitment and
consideration. If there were provocative incidents at the huts,
on the busses, for example, he simply refused to accept the
challenge. "I just made up my mind I wasn't going to no jail,"
he said. In brief, he would accept whatever he must, however
unjust, to stand by that promise to his bride.

At some point after the conclusion of the war—the exact
date had faded from his memory—he was finally able to obtain
one of the limited number of flattop houses allocated to
blacks. He remembered more clearly how frightened his wife
was in passing through the guarded gates to her new home.
They would later move to far better constructed housing in
Scarboro. Apparently, his longevity had allowed him to
resolve such resentment as he once may have had about his
early years at CEW. It was a valuable experience, he now
reminisced. "It'll learn you something. It'll learn you to
appreciate whatever you get."

Black Americans have been asked of recent years to forgive
much that has been perpetrated upon them by the white
majority. In general, they have somehow been willing to do
so. Such was the case in Oak Ridge. He and his wife were
typical in this regard. They were proud of their town. It was
not perfect. There was still some racial prejudice, but a
great deal of progress had been made. Indeed, Oak Ridge

integrated peacefully its public school grades 6 through 12 in
1955, one of the first Southern systems to so act. "She likes it
and I like it," he stated, "and we bought two lots to be buried
here." The couple, like most blacks who date their residence
to the war years, were atypical, however, of the larger
community who share the MED experience. They had
remained despite it. But the scars from that period
remained with them.[13] ·

Because Woodland and Scarboro were constructed
at approximately the same time, it was not uncommon in
the development period to make comparisons between them. Indeed at first glance, housing in both of
them appeared much the same—cinder-block,
single-family homes and frame duplexes. On closer
examination, however, significant distinctions were
apparent. Scarboro homes were essentially limited to
two bedrooms, while three and four were not uncommon in Woodland. Far fewer single-family homes,
as opposed to duplexes, were built in Scarboro than in
the other planned neighborhoods of Oak Ridge. Woodland was a case in point. Density of homes and
placement of houses on lots had apparently been a
greater consideration in development of the former
than the latter. Finally, for good or ill, attention given to
actual neighborhood design had obviously differed
substantially between the two locations. Scarboro was
characterized by a relatively simple layout. Cross
streets logically brought visitors back to the community's major perimeter roads or to the main artery
which led into the neighborhood from Illinois Avenue.
Woodland on the other hand, was a virtual jigsaw
puzzle of curving lanes, circles, and complex intercon-
necting streets. While the standard Oak Ridge alphabet system of street identification had been applied,
the layout was so complicated that it provided little
assistance to the unprepared. "Lost in Woodland," became the common response among Oak Ridgers for
anyone who temporarily could not be located.
Traveling again on Illinois, the incongruity so char-

acteristic of the city remained with the two historians right to the end. Beyond Shoney's, Hardee's, and K-Mart, just off Illinois on Scarboro Road, stood two imposing but now abandoned cement-block guard houses. Paint peeling, windows broken, and generally dilapidated, they had once been major check points for entry into the X-10 and Y-12 areas. As they approached the Clinch River and its Solway Bridge, the historians again sighted similar and equally dilapidated structures on Bethel Valley Road. They also had once provided security for X-10 and Y-12. Finally, the two stopped briefly, before crossing the river, to examine the original location of the reservation's Solway Gate. Subsequently, they made the return trip to Knoxville via the recently constructed Pellissippi Parkway, which cut in half the time that trip would have taken during the war.

MEMORY SIX
From the Pellissippi Parkway

The rough sketch of the epilogue was done. All that remained was to refine the notes into readable prose. That was not a major task. The conversation between the two historians gradually dwindled and virtually ceased. Each of them had become preoccupied with the same thought, which one finally put into words. Oak Ridge had taken much of their time and interest for almost four years. They had come to feel a very personal association with and affection for that city and its people. Now suddenly their professional involvement was at an end. It was a curious sensation.

What would become of Oak Ridge, they wondered aloud to each other. Was it destined at some point in the future to be just another East Tennessee town? Physically the city was in a constant state of change. It was simply a matter of time before only the most educated eye would be able to detect the Manhattan legacy which had set the city apart from other towns in the area, and, in truth, from every other town in the nation.

It was an unfortunate but nonetheless valid demographic fact of life that the MED generation would be gone in a number of years. Even among those Oak Ridgers interviewed by the historians about their wartime experiences at least two had since died. The children of that generation were now entering middle age and many had departed the city. Oak Ridge had, of course, also inherited large numbers of newcomers for whom the MED experience had no importance.

Yet for all these factors the historians concluded that, even if Oak Ridgers made an active effort, the status of ordinary town was not within the foreseeable future of that city. The 1945 flash in the sky at Hiroshima remained too clearly etched in the public record and popular memory. "Oak Ridge is where they made the bomb you know," one of the two remarked wryly to the other. Neither laughed.

Notes

CHAPTER ONE

1. The characterization is in large part drawn from Gordan Thomas and Morgan Witts, *Enola Gay* (New York, 1977), 8–9.

2. Ibid.

3. Dale E. Case, "Oak Ridge: A Geographic Study" (unpubl. diss., Univ. of Tennessee, 1955), 77–78; "Report on Proposed Site for Plant in Eastern Tennessee," no date, File MD 680.1, Box C-117-8, Central File Classified, Department of Energy, Oak Ridge, Tenn. (hereafter cited as CFCOR); "Oak Ridge: 20 Years After, Diversification Is the Goal," *Science* 150 (Nov. 12, 1965), 863; unpublished "Manhattan District History: Clinton Engineer Works" (hereafter cited as "Manhattan District History"), Book I, Vol. 12 (March 1947) Box 381638, Central File Unclassified, Federal Records Center, East Point, Ga. (hereafter cited as CFU, FRCEP).

4. George Robinson, *The Oak Ridge Story* (Kingsport, Tenn., 1950), 92.

5. "Basic Facts on the Oak Ridge Area," Sept. 15, 1948, File Basic Facts on Oak Ridge, Box 158540, CFU. FRCEP; M.J. Whitson to File, April 27, 1942, File CEW 600.3, Box 381641(16), CFU, FRCEP; Robinson, *The Oak Ridge Story*, 92.

6. C.C. Conklin to Area Engineer, Jan. 7, 1943, File MD 654, Box C-117-8, CFCOR; Case, "Oak Ridge: A Geographic Study," 134; John Poling to Officer in Charge, March 13, 1945, File ORBO, Box A-58-5(2), CFCOR.

7. Wilbur E. Kelly to Maj. Robert Blair, Jan. 3, 1943, File MD 624-Housing, Box C-117-8, CFCOR.

8. Ibid.

9. Stephane Groueff, *Manhattan Project* (Boston, 1967), 164.

10. Lt. D.C. Moore to File, Feb. 17, 1943, File MD 337, Box C-115-9, CFCOR; Interview #41, July 10, 1976.

11. Lenore Fine and Jesse Remington, *The United States Army in World War II: The Corps of Engineers; Construction in the United States* (Washington, D.C., 1972), 237.

12. Col. James C. Marshall to James C. White, Jan. 14, 1944, File MD 624, Box C-117-8, CFCOR.

13. Col. J.C. Marshall to File, Oct. 9, 1942, File General 624, Box 381638(13), CFU, FRCEP.

14. Fine and Remington, *The Corps of Engineers,* 669.

15. Theodore Rockwell, "Frontier Life Among the Atom Splitters," *Saturday Evening Post* 218 (Dec. 1, 1945), 29.

16. Interview #29, March 25, 1976.

17. "How Oak Ridge Street Program Grew," *American City* 63 (April 1948), 102–03.

18. Ibid.; Lt. Col. R.C. Blair to District Engineer, File CEW 600.05, Box 381641(16), CFU, FRCEP.

19. "The Town of Oak Ridge, 1946," File "History of Oak Ridge," Box 158547, CFU, FRCEP; "Manhattan District History," 8.2.

20. Thomas Boyers to P.S. Anderson, Aug. 25, 1944, File MD 333, Box C-117-8, CFCOR.

21. "Manhattan District History," 4.2, 4.3, 7.4, 7.12; Lt. Col. John S. Hodgson to R.H. Eichorn, April 26, 1945, File MAN 624 (R-A), Box 104870, CFU, FRCEP.

22. "Manhattan District History," 1.3; Robinson, *The Oak Ridge Story,* 45, 96. The peak figure of 100,000 includes 14,000 construction workers housed in trailers at K-25's so-called Happy Valley and who were not generally calculated in figures provided by the Corps on the population of "Oak Ridge."

23. Robinson, *The Oak Ridge Story,* 47, 96.

24. Interview #50, May 22, 1976.

25. "Manhattan District History," 7.1.

26. Capt. P.E. O'Meara to Lt. Col. J.S. Hodgson, May 22, 1944, File MAN 600.18 (R-A), Box 104870, CFU, FRCEP; Hayden Johnson, "Oak Ridge, Tennessee's Fifth Largest City," *Tennessee Planner* 6 (Dec. 1945), 67–74.

27. Capt. P.E. O'Meara, "Message from the Town Manager," *Oak Ridge Journal*, Oct. 2, 1943.
28. T.C. Williams to Area Engineer, Nov. 20, 1943, File Tracts: Demolition, Box 158527, CFU, FRCEP.
29. Maj. L.W. Devereaux to Lt. Col. R.C. Blair, April 16, 1943, File CEW 600.05, Box 38641(16), CFU, FRCEP.
30. C.N. Hernandez to Maj. E.J. Bloch, Aug. 19, 1944, File MAN 600.12, Box 104870(19), CFU, FRCEP.
31. Lt. Col. J.S. Hodgson to C.N. Hernandez, Sept. 19, 1944, File CEW 600.05, Box 381641(16), CFU, FRCEP.
32. W. Norris Wentworth to H.S. Russell, Jan. 29, 1945; Maj. M.O. Swanson to Maj. M.M. Pettyjohn, March 28, 1945, File CEW 600.05, Box 381641(16), CFU, FRCEP.
33. Typescript Speech to Oak Ridge Kiwanis Club (speaker unidentified), Oct. 7, 1946, File MAN 080, Box 141909, CFU, FRCEP.
34. Lt. Col. T.T. Crenshaw to Chairman, Federal Communications Commission, April 15, 1944, File Western Union Operation, Box 158528, CFU, FRCEP.

CHAPTER TWO

1. Robinson, *The Oak Ridge Story*, 17–19.
2. Fine and Remington, *The Corps of Engineers*, 655; "United States versus 52,600 Acres of Land in Roane and Anderson Counties, Tennessee and Ed Browder, et al.," Oct. 6, 1942, Civil Action No. 429, District Court of the United States, Eastern District of Tennessee, Northern Division, Knoxville. (The total land taken ultimately was 56,000 acres.)
3. A. S. Landry to Marguerite Owen, March 9, 1943, Office of the Engineers, TVA Library, Knoxville.
4. Robinson, *The Oak Ridge Story*, 35–36.
5. Fine and Remington, *The Corps of Engineers*, 663.
6. Secretary of War to Congressman John J. Jennings, Aug. 11, 1945, File MAN 032, Box 104873, CFU, FRCEP; Interview #20, July 29, 1976.
7. Robinson, *The Oak Ridge Story*, 30.
8. Interview #20, July 29, 1976.
9. Case, "Oak Ridge: A Geographic Study," 30–38.
10. Ibid.

11. Ibid.; *Knoxville News-Sentinel,* July 23, 1943.

12. Gen. Leslie R. Groves, *Now It Can Be Told* (New York, 1962), 26–27; Capt. George B. Leonard to Deputy District Engineer, July 13, 1943, File CEW 680.2, Box 381643, CFU, FRCEP.

13. *Sixteenth Census of the United States: 1940 Population,* Vol. II, Pt. 6, "Characteristics of the Population," Dept. of Commerce, Bureau of the Census (Washington, D.C., 1943), 590; Robinson, *The Oak Ridge Story,* 47.

14. Interview #45, March 27, 1976. While a significant minority, especially of scientific personnel, came from other areas of the country, the majority, 65 percent, came from "the states comprising the great Tennessee Valley and adjoining states in the South"; Robinson, *The Oak Ridge Story,* 47.

15. Fine and Remington, *The Corps of Engineers,* 673; Richard Hewlett and Oscar E. Anderson, *The New World: A History of the United States Atomic Energy Commission* (University Park, Penn., 1962), 118–19.

16. Robinson, *The Oak Ridge Story,* 74.

17. *Clinton* (Tenn.) *Courier-News,* July 1, 1943; Interview #20, July 29, 1976; Maj. Raymond Welch to District Engineer, July 19, 1943, File 230.14, Box C–118–7 (2); Roane-Anderson-CEW Payroll, Oct. 7, 1945, File MD 248, Box C-117-5(2), CFCOR.

18. Maj. E.J. Bloch to Capt. L.D. Geiger, Feb. 12, 1945, File MD 624, Box C-117-8, CFCOR; "City That Atoms Built," *Business Week,* Oct. 27, 1945, p. 24; "Summary of Discussion on Housing," Nov. 3, 1943, File CEW 624, Box 381642, CFU, FRCEP.

19. *Clinton Courier-News,* Feb. 17, 1944; Maj. Gen. Leslie Groves to Chief, Federal Agencies Liaison Branch, Sept. 6, 1943, File 671.2, Box C-117–5(2), CFCOR.

20. *Clinton Courier-News,* Nov. 1, 1943.

21. Ibid., July 15, 1943.

22. Fine and Remington, *The Corps of Engineers,* 673; *Clinton, Courier-News,* March 4, 1943; Interview #15, May 15, 1976.

23. First Lt. Nicolas Del Genio to Officer in Charge, May 18, 1945, File MD 333, Box C-117-8(2), CFCOR.

24. Federal Security Agency, Knoxville-Clinton Progress Report II, Oct. 14, 1944, File MD 624, Box C-117-8, CFCOR.

25. Lt. Col. Thomas Crenshaw to Capt. P.E. O'Meara, May

8, 1944, File Tracts, Farm Equipment, Box 158527, CFU, FRCEP.

26. Congressman John Jennings to Secretary of War, July 26, 1945; Secretary of War to Congressman John Jennings, Aug. 11, 1945 both in File MAN 032 (Jennings Jr.) Box 104873, CFU, FRCEP.

27. L.V. Denton and B.P. Hartman to Sen. Tom Stewart, May 13, 1946; Maj. Gen. Leslie Groves to Sen. Tom Stewart, May 24, 1946 both in File MAN 032 (Stewart), Box 104873, CFU, FRCEP.

28. Ibid.

29. Interview #20, July 29, 1976; Capt. Victor A. Sheridan to War Production Board, Consumer Goods Section, June 6, 1945, File Miller's Dept. Store, Box A-58-5, CFCOR.

30. Federal Security Agency, Knoxville-Clinton Progress Report II, Oct. 14, 1944, File MD 624, Box C-117-8, CFCOR; *Clinton Courier-News*, Jan. 6, 1944.

31. Ibid.; William A. Kelly, "The State and County Government Relationships of Oak Ridge, Tennessee," (Unpubl. M.A. Thesis, Univ. of Tennessee, 1951), 64; *Knoxville News-Sentinel*, Aug. 18, 1943; Maj. Gen. Leslie Groves to Sen. Kenneth McKellar, April 9, 1946, File MAN 032 (McKellar), Box 104873, CFU, FRCEP.

32. *Clinton Courier-News*, Aug. 25, Nov. 25, 1943; Jan. 20, June 18, 1944; *Knoxville News Sentinel*, Aug. 10, 1944; Record of Conversation, Col. C.D. Cornell, Lt. Col. J. Weil, and Capt. H. Traynor, 1945 (Precise date not given), File MD 337, Box C-115-9, CFCOR.

33. Lt. Col. T.T. Crenshaw to Stone and Webster Engineering Co., Feb. 19, 1944; Maj. E.J. Bloch to Roane-Anderson Co., May 26, 1944, both in File MAN 611, Box 104870, CFU, FRCEP; Col. E.H. Marsden to Maj. Gen. Leslie Groves, March 13, 1944, File CEW 600.93, Box 381641; *Oak Ridge Journal*, Jan. 25, 1945.

34. *Clinton Courier-News*, Dec. 16, 1943.

35. Ibid.; Oct. 12, 1945; Kelly, "The State and County Governmental Relations of Oak Ridge," 27.

CHAPTER THREE

1. Robinson, *The Oak Ridge Story*, 93–94; "Manhattan District History," 6.1.

2. Ibid.; "The City That Atoms Built," 21–28; *Oak Ridge Journal*, May 10, 1945.

3. *Oak Ridge Journal*, Jan. 22, 1944.

4. Clinton Hernandez to Area Engineer, Oct. 20, 1943, File R-A General, Box 158529; Capt. P.E. O'Meara to Cost Section, Dec. 29, 1943, File MAN 600.18, Box 104870, both in CFU, FRCEP.

5. Capt. E.J. Bloch to T.C. Williams, Nov. 19, 1943, File MAN 600.18 Box 104870, CFU, FRCEP.

6. "The City That Atoms Built," 26; Recapitulations of Concessions, Nov. 30, 1944, File Concessions Granted, Box 158528, CFU, FRCEP; Lt. Col. T.T. Crenshaw to Stone and Webster Corp., Jan. 26, 1944, File CEW 532, Box 381640, CFU, FRCEP.

7. Robinson, *The Oak Ridge Story*, 94.

8. Capt. P.E. O'Meara to Lt. Col. J.S. Hodgson, May 22, 1944, File MAN 600.18, Box 104870, CFU, FRCEP.

9. Ibid.

10. "Minutes of the Central Facilities Advisory Council Meeting," April 1944, File CFCA, Box A-58-5 (cont.); Roane-Anderson, "Report of Operations, July 1944," Box A-58-5, CFCOR.

11. Ibid.

12. Ibid.; L.J. Cotton to Lt. Col. Vanden Bulck, Dec. 31, 1943, File General Correspondence, Recreation and Welfare Association, Box 158597, CFU, FRCEP.

13. "The City That Atoms Built," 21.

14. Roane-Anderson, "Report of Operations, July 1944"; Lt. Col. J.S. Hodgson to C.N. Hernandez, Oct. 6, 1944, File MAN 624 (R-A), Box 104870, CFU-FRCEP.

15. Roane-Anderson "Report of Operations, July 1944"; Capt. P.E. O'Meara to C.N. Hernandez, Dec. 15, 1943, File MAN 624 (R-A); O'Meara to Hernandez, April 14, 1944, File (TE) 624, both in Box 104872, CFU-FRCEP; "Manhattan District History," 16.5.

16. Lt. Col. J.S. Hodgson to C.N. Hernandez, Oct. 6, 1944; Major E.J. Bloch to C.N. Hernandez, Nov. 4, 1945, both in File MAN 624 (R-A), Box 104870, CFU FRCEP.

17. Maj. E.J. Bloch to A.A. Carswell, Jan. 29, 1945, File MAN (R-A), Box 104870, CFU, FRCEP.

18. Minutes of the Central Facilities Advisory Council, Nov. 13, 1944, File CFAC, Box A-58-5(cont.) (2), CFCOR.

19. Maj. W.E. Kelly to Lt. Col. T.T. Crenshaw, Sept. 10, 1943; James C. White to Col. K.D. Nichols, Oct. 25, 1943, File MD 624; Maj. W.E. Kelly to File, Jan. 15, 1944, File MD 337, Box C-117-8, CFCOR.

20. C.E. Center to Lt. Col. J.C. Stowers, Jan. 19, 1944, File 624, Box C-117-8, CFCOR.

21. Lt. Col. J.S. Hodgson to C.N. Hernandez, July 17, 1944, File MAN 624 (R-A), Box 104870; Lt. Col. J.S. Hodgson to Lee G. Warren, Sept. 25, 1946, File 624 Housing (T-E), Box 104872, CFU, FRCEP.

22. Lt. Col. J.S. Hodgson to C.N. Hernandez, July 17, 1944, File MAN 624 (R-A), Box 104870, CFU, FRCEP.

23. Capt. P.E. O'Meara to C.N. Hernandez, Dec. 15, 1943, File MAN 624 (R-A); O'Meara to Hernandez, April 14, 1944, File MAN 624 (T-E), both in Box 104872, CFU, FRCEP.

24. Lt. Col. J.S. Hodgson to C.N. Hernandez, Oct. 6, 1944; Maj. E.J. Bloch to C.N. Hernandez, May 4, 1945, both in File MAN 624 (R-A), Box 104870, CFU, FRCEP.

25. Maj. E.J. Bloch to A.A. Carswell, Jan. 29, 1945, File MAN 624 (R-A), Box 104870, CFU, FRCEP.

26. Minutes of the Central Facilities Advisory Council, April 13, 1944, File CFAC, Box A-58-5(cont.) (2), CFCOR.

27. Ibid.

28. Lt. Col. J.S. Hodgson to C.N. Hernandez, July 17, 1944, File MAN 624 (R-A), Box 104870, CFU, FRCEP.

29. Roane-Anderson, "Report of Operations, July 1944."

30. "Manhattan District History," 7.8.

31. Capt. P.E. O'Meara to C.N. Hernandez, Jan. 12, 1944, File MAN 600.18 (R-A), Box 104870, CFU, FRCEP; Ivy Lee and T.J. Ross, "A Public Relations Study of the CEW Project at Oak Ridge, Tennessee, December, 1944–January, 1945," Lee-Ross Public Relations Study, Box 158530, CFU, FRCEP; the latter hereafter cited as Lee-Ross Study, 1944–45.

32. Capt. P.E. O'Meara to Lt. Col. J.S. Hodgson, May 22, 1944, File MAN 600.18 (R-A), Box 104870, CFU, FRCEP.

33. Roane-Anderson, "Report of Operations, July 1944."

34. Ibid.; L.J. Cotton to Lt. Col. Vanden Bulck, Dec. 31, 1943, File General Correspondence, Recreation and Welfare Association, Box 104872, CFU, FRCEP; "Manhattan District History," 8.6.

35. Fred Ford and Fred Peitzsch, "A City is Born," 9–10 (this unpublished and unofficial history of Oak Ridge from

its inception to approximately 1960 was written by two employees at the AEC at the request of the agency; eighteen copies were subsequently produced by lithograph process); "The City That Atoms Built," 26; Recapitulation of Concessions, Nov. 30, 1944, File Concessions Granted, Box 158528, CFU, FRCEP.

36. C.M. Winfrey to A&P Tea Company, Aug. 27, 1946, File CEW 680.48, Box 14917, CFU, FRCEP.

37. Thomas Boyers, OPA to Capt. P.S. Anderson, Aug. 25, 1944; Maj. H.S. Traynor, Memo for Manhattan District History, Jan. 2, 1945, File MD 333, both in Box 141909, CFU, FRCEP.

38. Report by C.N. Hernandez to District Engineer, May 14, 1944, File Roane-Anderson Report of Operations, Box 104868; Lt. Col. T.T. Crenshaw to Col. K.D. Nichols, May 10, 1945, File CEW 680.314, Box 381643, CFU, FRCEP; "The City That Atoms Built," 26.

39. Roane-Anderson, "Report of Operations, July 1944."

40. *Oak Ridge Journal,* May 10, 1945; "The City That Atoms Built," 24; Roane-Anderson Weekly Payroll, Nov. 5, 1944, File MD 248, Box 104868, CFU, FRCEP.

41. Robinson, *The Oak Ridge Story,* 97.

42. L.A. Waddel to District Engineer, Sept. 27, 1945, File MAN 330.15 (CEW), Box 104884, CFU, FRCEP.

43. Ibid.

CHAPTER FOUR

1. Maj. Curtis Nelson to Capt. W.A. Fogg, Sept. 17, 1943; Maj. Curtis Nelson to Area Engineer, Oct. 16, 1943, File Study of Housing Policy, Box C-117-8, CFCOR.

2. Col. K.D. Nichols to All Concerned, Nov. 9, 1943; Lt. Col. E.H. Marsden to Area Engineer, Jan. 24, 1944; Lt. Col. E.H. Marsden to Capt. W.A. Fogg, Jan. 31, 1944, File Study of Housing Policy, Box C-117-8, CFCOR.

3. Samuel Baxter to Capt. P.E. O'Meara, June 13, 1944, No File, Box A-2-6; Lt. Col. E.H. Marsden to Area Engineer, Jan. 24, 1944; Col. K.D. Nichols to File, Nov. 9, 1943, File Study of Housing Policy; Minutes of the Central Facilities Advisory Council Meeting, Jan. 23, 1945, File CFAC Minutes, Box C-117-8, CFCOR.

4. Col. K.D. Nichols to All Concerned, May 6, 1944, File

Study of Housing Policy; Minutes of the Central Facilities Advisory Council Meeting, Jan. 23, 1945, File CFAC Minutes, Box C-117-8, CFCOR; Maj. W.T. Kelly to Lee G. Warren, Feb. 29, 1944, File (TE) 624, Box 104872, CFU, FRCEP.

5. Lt. Col. Curtis Nelson to District Engineer, Nov. 15, 1944, File MD 330; N.H. Smith to Lt. Col. J.S. Hodgson, Dec. 12, 1944, File MD 621, Box C-117-8; Capt. W.A. Barger to District Engineer, March 2, 1945, File MD 320.2, Box C-118-1, CFCOR.

6. "Reservation Population, 1944-45," File Reservation Population, Box A-58-5, CFCOR.

7. A.L. Baker to Col. J.C. Stowers, July 14, 1944, File MD 624 Housing, Box C-117-8, CFCOR.

8. Col. K.D. Nichols to Maj. Gen. L.R. Groves, Dec. 7, 1944, File MD 624 Housing, Box C-117-8, CFCOR.

9. Clark Center to District Engineer, Nov. 19, 1946, File MAN 624 (CCCC) Housing, Box 104895, CFU, FRCEP.

10. *Oak Ridge Journal*, Oct. 23, 1943, May 25, 1944; Lt. Col. J.S. Hodgson to C.N. Hernandez, June 17, 1944, File MAN 618.35, Box 104870 (19); Col. Philip Kromer to A.J. Brock, June 18, 1946, File Policy, Box 158595, CFU, FRCEP.

11. Dr. Eric Clarke to Col. C. Warren, July 6, 1945, File MD 701, Box 117-9 (cont.), CFCOR.

12. Dr. Eric Clarke to Maj. Charles E. Rea, Oct. 12, 1944, File MAN 231.23, Box 134303, CFU, FRCEP.

13. Roane-Anderson, "Report of Operations, July 1944."

14. Ibid.; Minutes of the Central Facilities Advisory Council, April 6, 1944, File Study of Housing Policy, Box A-58-6 (2), CFCOR.

15. Roane-Anderson, "Report of Operations, July 1944"; Lt. Col. T.T. Crenshaw to District Engineer, Nov. 2, 1943, File CEW 624, Box 381642, CFU, FRCEP.

16. Ibid.

17. Capt. E.J. Bloch to Stone and Webster, Feb. 23, 1944, File CEW 624, Box 381642; Capt. P.E. O'Meara to C.N. Hernandez, April 6, 1944, File Central Facilities Advisory Council Meeting, Box 158537, CFU, FRCEP.

18. Colored Camp Council to Col. J.S. Hodgson, July 21, 1944, File CEW 624, Box 381642, CFU, FRCEP.

19. Ibid. The reference to "white family huts" is an incorrect one and probably refers to the Victory Cottages. No white families lived in hutments during the war.

20. Lee-Ross Study, 1944–45.

21. Maj. E.J. Bloch, Minutes of the Recreation and Welfare Council, Nov. 1, 1944, File Major Bloch Minutes, Box A-58-4, CFCOR; Swep T. Davis to District Engineer, June 27, 1946, File 631 R&W, Box 104887, CFU, FRCEP.

22. Minutes of the Recreation and Welfare Council, Sept. 26, 1945, File R&WA Meetings, Box 158597; Rev. Beauford Bradshaw to District Engineer, July 9, 1946, File Churches: Inter-Denominational, Box 158536, CFU, FRCEP.

23. Capt. T.W. Taylor memo to the files, March 14, 1945, File Churches: Baptist, Box 158536, CFU, FRCEP.

24. Lee-Ross Study, 1944-45.

25. Conclusions are based on extensive examination of reservation police records for the war period. Unfortunately, these records have since been destroyed by the Energy Research and Development Agency.

26. Case 474, Stone and Webster badge 32460, Aug. 27, 1944, Oak Ridge Police Records, now destroyed.

27. Case 77, Stone and Webster badge 1696, Aug. 9, 1944, Oak Ridge Police Records, now destroyed.

28. Minutes of the Recreation and Welfare Council, June 13, 1945, File R&WA Meetings, Box 158597, CFU, FRCEP; *Oak Ridge Journal*, Feb. 8, 1945.

29. Case 13325, Dec. 9, 1945, Oak Ridge Police Records, now destroyed.

30. *Oak Ridge Journal*, Nov. 9, 1944.

31. M.R. Cook to C.L. Barnard, June 3, 1944, File MD 631, Box 117-9, CFCOR.

32. Swep T. Davis to Lt. Col. J.S. Hodgson, Dec. 30, 1944, File CEW 230.65, Box 381640, CFU, FRCEP.

33. Minutes of a Special Meeting of the Council, Recreation and Welfare Association, March 30, 31, April 1, 1944, File MD 334, Box C-117-8 (cont.), CFCOR; Lt. Col. J.S. Hodgson to K.H. Kline, May 9, 1945, File CEW 000.17, Box 381639, CFU, FRCEP.

34. *Oak Ridge Journal*, Oct. 19, 1944; George Chandler to Roane-Anderson Company, Aug. 13, 1945, File MAN 000.7 (CEW), Box 104883, CFU, FRCEP.

35. "The Town of Oak Ridge," 1946, File History of Oak Ridge Film Draft, Box 158547, CFU, FRCEP.

36. *Ibid.*

37. *Oak Ridge Journal*, Oct. 2, 1943.
38. Minutes of the Central Facilities Advisory Council (Housing and Townsite), June 8, 1944, File MD 314.7, Box C-117-9 (2), CFCOR.
39. Minutes of the Central Facilities Advisory Council, June 1, 1944, File Study of Housing Policy, Box C-117-8, CFCOR; Rev. J.C. Sieck to Capt. W.T. Taylor, March 30, 1945, File Churches, Box 158536, CFU, FRCEP.
40. Capt. G.L. Ryan to Lt. Col. J.S. Hodgson, Aug. 28, 1944, File Church Attendance Survey, Box A-58-4 (cont.), CFCOR.
41. Lt. Col. J.S. Hodgson to Rev. B.M. Larsen, Aug. 11, 1945, File CEW 000.3, Box 158532, CFU, FRCEP.
42. Maj. Charles E. Rea to H.G. Hoberg, Sept. 20, 1944, File MD 314.7, Box C-115-6 (1), CFCOR.
43. Stafford L. Warren to District Engineer, Feb. 15, 1944, File MD 700, Box C-117-9 (cont.), CFCOR; Capt. Fred Houston to Recreation and Welfare Association, Sept. 16, 1944, File CEW 230.65, Box 381640, CFU, FRCEP.
44. Robinson, *The Oak Ridge Story*, 65; Stafford L. Warren to Maj. Gen. L.R. Groves, Oct. 20, 1944, File MD 726.1, Box C-115-4, CFCOR.
45. Col. Philip Kromer to Mrs. Sarah Cater, May 29, 1946, File MAN 680.48, Box 141918 (86); Police Report, Feb. 28, 1947, File Case Reports, Box 158537 (11), CFU, FRCEP.
46. Swep T. Davis to District Engineer, Dec. 19, 1944, File CEW 230.65; Maj. R.F. Looney to Ford, Bacon and Davis, Inc., Jan. 8, 1945, File CEW 532, Box 381640; unsigned memo re James J. Girder family, Fall 1946, File MAN 330.14, Box 141911, CFU, FRCEP.
47. Lee-Ross Study, 1944-45.
48. Theodore Rockwell, "Frontier Life Among the Atom Splitters," *Saturday Evening Post* 218 (Dec. 1, 1945), 45.
49. Fine and Remington, *The Corps of Engineers*, 661.

CHAPTER FIVE

1. Robinson, *The Oak Ridge Story*, 67.
2. Notebook revisions for "A City is Born" (in the hands of the Dept. of Energy, Oak Ridge), 41.
3. R.J. McLeod to W.B. Parsons, Sept. 18, 1945, File MD

Miscellaneous, Box 158535, CFU, FRCEP; Lt. Col. W.B. Parsons to Lt. Col. John Lansdale, March 19, 1945, File ORBO, Box A-58-5(cont.), CFCOR.

4. Maj. R.J. McLeod Memo for Record, Aug. 24, 1945, File MD Miscellaneous, Box 158535, CFU, FRCEP.

5. Maj. R.J. McLeod to Walter Stagg, Feb. 26, 1945, File ORBO, Box A-58-5 (cont.), CFCOR; Maj. L.R. Block to Lt. Col. W.B. Parsons, Sept. 7, 1944, File CEW Guard Force, Box 158533 (7), CFU, FRCEP; Lt. William Foltz to Capt. E.B. Brown, May 19, 1945, File ORBO 200.07, CFCOR.

6. Director, Security Division AEC, to Manager, Field Operations, Feb. 7, 1947, File MAN 572.1, Box A-58-5 (cont.), CFCOR; Col. K.D. Nichols to Maj. A.C. Johnson, Aug. 26, 1943, File CEW 600.96, Box 381641, CFU, FRCEP.

7. A.G. Underwood to W.J. Hatfield, Aug. 2, 1944, File Jones 230.146, Box 104866; L.P. Riordan to Maj. L.R. Block, Dec. 21, 1944, File MAN 230.146, Box 104861; Maj. James Shackelford to Roane-Anderson Company, April 10, 1944, File Roane-Anderson: General, Box 158529; J.E. Brock to Maj. Walter W. Stagg, April 9, 1945, File MAN 230.146 (R-A) Blacklisted Employees, Box 104869 (18), all in CFU, FRCEP.

8. Maj. L.R. Block to Ford, Bacon and Davis Inc., April 15, 1944, File Ford, Bacon and Davis 680.2, Box 104864, CFU, FRCEP.

9. Capt. Samuel Baxter to Maj. L.R. Block, June 13, 1944, File CEW Security Policy, 1944-45, Box A-58-5 (cont.) (2), CFCOR.

10. Minutes of the Recreation and Welfare Association Council, Oct. 11, 1944, File Minutes of the Recreation and Welfare Association Council, Box A-58-5(cont.) (2), CFCOR; Lt. Col. J.S. Hodgson to Tri-State League Professional Baseball, Oct. 19, 1945, File MAN 080, Box 104883; Lt. Col. J.S. Hodgson to Oak Ridge Recreation and Welfare Association, July 25, 1944, File CEW 230.65, Box 384160, CFU, FRCEP.

11. Maj. W.T. Sergeant to File, March 23, 1946, File Churches: Nazarrene, Box 158536, CFU, FRCEP.

12. Capt. Leo T. Zbanek to Roane-Anderson Company, Oct. 9, 1945, File RA 600.96, Box 104887, CFU, FRCEP.

13. Capt. James Haley to Lt. Col. W.B. Parsons, May 29, 1944, File CEW Security Policy, 1944-45, Box A-58-5(cont.) (2), CFCOR.

14. Minutes of the Central Facilities Advisory Council

Meeting April 13, 1944, File CFAC, Box A-58-5 (cont.); Report by C.N. Hernandez to District Engineer, Aug. 3, 1944, File Report of Operations, Box A-58-4, both in CFCOR.

15. Robinson, *The Oak Ridge Story*, 68.

16. Ibid., 69.

17. Ibid.

18. Ibid.

19. Ibid., 70.

20. Maj. H.G. Hoberg to C.N. Hernandez, Dec. 16, 1944, File RA 200.2, Box 104869, CFU, FRCEP; *Oak Ridge Journal*, Dec. 7, 1944.

21. Agent COB report, April 11, 1945, File ORBO, Box A-58-5 (cont.) (2), CFCOR.

22. Col. Philip Kromer to Charles L. Thompson, Oct. 3, 1946, File MD 333, Box C-117-8, CFCOR; Case 3791, Feb. 6, 1945, Oak Ridge Police Records, now destroyed.

23. H.S. Trayner to File, July 1, 1944, File MD 370.61, Box A-47-1, CFCOR.

24. Maj. Paul Rossell to Lt. Col. T.T. Crenshaw, April 4, 1944, File Effects on CEW of a Failure at Norris Dam, Box C-117-10, CFCOR.

25. Col. K.D. Nichols to Maj. Gen. L.R. Groves, July 13, 1944, File MD 314.7, Box C-115-2, CFCOR.

26. One such bit of evidence became apparent at a talk given by one of the authors to "The 43 Club" in Oak Ridge in December 1977. Attendance was approximately 100 persons of whom roughly 20 percent admitted serving in this capacity. Interview data obtained by the authors from MED-period residents also suggests that a substantial number of residents were intelligence operatives.

27. *Information Bulletin for Oak Ridge Residents* (Roane-Anderson, June 15, 1944), 1.

28. Rev. B.M. Larsen to Capt. T.W. Taylor, Jan. 2, 1945, File United Church, Box 158532, CFU, FRCEP.

29. Capt. W.T. Sergeant, CEW General Order 30, May 7, 1945, File Special Orders, Box 158543, CFU, FRCEP.

30. Lt. Col. P. Cornelius to J.A. Jones Construction Company, Dec. 7, 1944, File Jones 200.5, Box 104866, CFU, FRCEP.

31. While perhaps apocryphal, this story is widely accepted as true and is frequently repeated by present-day Oak Ridgers.

32. Interview #33, June 26, 1976.

33. William Manchester, *The Glory and the Dream: A Narrative History of America, 1932–72* (Boston, 1973), 375.

34. Groves, *Now It Can Be Told*, 101.

35. Plant Superintendent, Carbide and Carbon Corporation, to District Engineer, March 9, 1945, File MD 381, Box A-55-5; Capt. W.F. Cronin to Lt. Col. Thomas Crenshaw, April 19, 1945, File MD 381, Box A-55-7 (2), both in CFCOR; Capt. T.W. Taylor to Dr. A.H. Blankenship, File Housing, Box 158528, CFU, FRCEP; *Oak Ridge Journal*, April 26, July 26, 1945.

36. Robinson, *The Oak Ridge Story*, 101–104.

37. Manchester, *The Glory and the Dream*, 378–79.

38. Arthur Holly Compton to Maj. Gen. L.R. Groves, July 10, 1945, File MD 337, C-115-9 (2), CFCOR. Compton's memo is a summary of earlier correspondence between the two and in response to Groves' opposition to Compton's proposals. The earlier correspondence is not in the file. Here Compton accedes to Groves' prohibition of formal meetings and, instead, suggests less formal discussion among individuals.

39. Farrington Daniels to Arthur Holly Compton, July 12, 1945, File MD 337, C-115-7 (cont.), CFCOR; Arthur Holly Compton to Capt. J.H. McKinley, July 17, 1945, File MD 337, Box C-117-4 (2), CFCOR.

40. Interviews #50, May 15, 1976, and #41, May 1, 1976.

41. Groueff, *Manhattan Project*, 359.

42. Manchester, *The Glory and the Dream*, 380.

43. Groves, *Now It Can Be Told*, 317, 319; Groueff, *Manhattan Project*, 362.

44. Manchester, *The Glory and the Dream*, 381–82.

45. Robinson, *The Oak Ridge Story*, 106.

46. Manchester, *The Glory and the Dream*, 382.

47. Ibid.

48. Douglas Plywood Association to Clinton Engineer Works, Aug. 7, 1945, File MAN 000.7, Box 104873 (22), CFCOR.

49. American Lanudry Machinery Company to Col. K.D. Nichols, Aug. 9, 1945, File MAN 000.7, Box 104873 (22), CFU, FRCEP.

50. Maj. W.A. Rathery to White, Louis, Rostetter and Sons

Company, Aug. 27, 1945, File MAN 000.7, Box 104873(22), CFU, FRCEP.

51. Lt. G.O. Robinson to Tucker Wayne Co., Sept. 13, 1945, File MAN 000.7, Box 104873 (22), CFU, FRCEP.

52. Professor Herbert S. Harned to Maj. E.J. Bloch, Sept. 20, 1945, File MAN 000.7, Box 104873 (22), CFU, FRCEP.

53. Robinson, *The Oak Ridge Story*, 106.

54. Ibid., 106–107. The interviews conducted by the authors with a number of individuals who were in Oak Ridge on August 6 essentially verify the scene as described by Robinson.

55. Ibid.

56. Interview #29, March 25, 1976.

57. Interview #10, March 25, 1976.

58. Lt. Col. P.T. Preston to Col. K.D. Nichols, Aug. 15, 1945, File MAN 200.73, Box 104873 (22), CFU, FRCEP.

59. Robinson, *The Oak Ridge Story*, 108.

CHAPTER SIX

1. Press release, Col. K.D. Nichols, Sept. 6, 1945, File MAN 000.7, Box 104873(22), CFU, FRCEP.

2. Maj. W.C. Young, Jr. to Col. P.F. Kromer, May 30, 1946, File Statistical Data, Box 158537; G.O. Robinson to G. Paul Crowder, Sept. 11, 1946, File MAN 000.77, Box 141909; Capt. William A. Bonnet to J. G. LeSieur, July 16, 1946, File CEW 230, Box 156205, CFU, FRCEP; Cabell Phillips, "Oak Ridge Ponders its Clouded Future," *New York Times Magazine,* April 13, 1947, pp. 12–13; Robinson, *The Oak Ridge Story,* 49.

3. "Oak Ridge—Reconverted," *Business Week* (Aug. 3, 1946), 19.

4. Ibid.; "Oak Ridge: Life Where the Bomb Begins," *Newsweek* 28 (Aug. 5, 1946), 32–4; Warren Ogden, "The A-Bomb's Home," *New York Times Magazine,* April 14, 1946, p. 16.

5. "Oak Ridge—Reconverted," 19.

6. C. Rufus Rorem, bound report, "The Oak Ridge Hospital," Aug. 1, 1948, File Oak Ridge Hospital, Box 158581; Report, "Changes in the Oak Ridge Population," Jan. 15, 1948, File Changes in Population, Box 158602, both in CFU, FRCEP.

7. Report, "Reservation Population," June 15, 1946, File 019.4, Box 104883; Col. P.F. Kromer to Rep. J. Percy Priest, Aug. 28, 1946, File MAN 032, Box 104903, both in CFU, FRCEP.

8. W. Norris Wentworth to Capt. Leo T. Zbanek, Aug. 14, 1945, File MAN 624 (R-A), Box 104870; Lt. Col. J.S. Hodgson to Sen. Tom Stewart, Nov. 19, 1945, File MAN 032, Box 104873 (22); F.W. Cook to District Engineer, Oct. 8, 1945, Capt. Leo Zbanek to Canteen Food Services, Oct. 17, 1945, File 624, Box 104873; Roane-Anderson Company to Occupants of the White Hutment Camp, Oct. 11, 1945, File Trailer Camps, Box 158528, all in CFU, FRCEP.

9. Lt. Col. J.S. Hodgson to File, Oct. 11, 1945, File MAN 334, Box 104884, CFU, FRCEP.

10. Lt. Col. J.S. Hodgson to File, Nov. 8, 1945, File MAN 334, Box 104884; F.W. Cook to District Engineer, Dec. 29, 1945, File MAN 624 (R-A), Box 104887; T.R. Disbrow to District Engineer, Dec. 24, 1946, File MAN 680.48, Box 141918; Maj. William A. Bonnet to File, Jan. 17, 1946, File MAN 334, Box 104884 (33), all in CFU, FRCEP.

11. Metz T. Lochard to G.O. Robinson, Dec. 31, 1945, File MAN 330.13, Box 104882, CFU, FRCEP; on press coverage see for example, "Oak Ridge: Life Where the Bomb Begins," 32-34.

12. Scott E. Williamson to Congressman John Jennings, Jan. 14, 1946, File MAN 032, Box 104873, CFU, FRCEP.

13. Capt. Leo T. Zbanek to Roane-Anderson, Jan. 3, 1946, File Concessionaires Gross Receipts, Box A-58-5 (cont.), CFCOR; Richard D. Sheahan to District Engineer, Nov. 29, 1945, File General Correspondence, Recreation and Welfare Association, Box 158597 (71), CFU, FRCEP.

14. "Oak Ridge: Life Where the Bomb Begins," 33–34.

15. Leslie Carr to Col. K.D. Nichols, April 10, 1945, File CC080, Box 104861; Lt. Col. Curtis Nelson to National Labor Relations Board, June 27, 1946, File MAN 080, Box 104883, both in CFU, FRCEP.

16. "Policy on Union Meetings," Jan. 27, 1945, File Policy on Union Activity, Box A-58-5 (cont.) (2), CFCOR; Lt. Col. Curtis Nelson to Col. P.F. Kromer, July 16,1946, File Labor, Box 158530, CFU, FRCEP.

17. See for example: Report from Confidential Informant 41K, March 5, 1945; Lee Porter, Special Agent CIC to

Officer in Charge, Jan. 14, 1946; Report from Confidential
Informant 58-K, March 11, 1946; Report from Confidential
Informant 800-A, March 12, 1946; Lt. Max E. Rogers to Col.
D.F. Shaw, June 12, 1946, all in File Labor Activities, Box
A-58-5(cont.)(2), CFCOR; Dewey R. Davis and P.H. Ton-
guette to District Engineer, June 17, 1946, File MAN
330.14(CEW), Box 104884, CFU, FRCEP.
18. Maj. Gen. L.R. Groves to District Engineer, Aug. 30,
1946, File Labor, Box 158530, CFU, FRCEP.
19. Col. P.F. Kromer to Lt. Col. D.G. Williams, Sept. 19,
1946, File Labor Relations, Box 158543, CFU, FRCEP.
20. Lt. Col. D.G. Williams to Col. P.F. Kromer, May 14,
1946; Lt. Col. D.F. Shaw to Col. E.E. Kirkpatrick, June 4,
1946, both in File Picketing, Box A-58-5 (cont.)(2), CFCOR.
21. First Lt. William Cates to Maj. R.J. McLeod, Feb. 20,
1946; 1st Lt. A.J. Anderson to Col. D.F. Shaw, April 15, 1946,
both in File Labor Activities, Box A-58-5 (cont.)(2), CFCOR;
Telegram, Maj. R.J. Coakley to Lt. Col. Curtis A. Nelson, Aug.
6, 1946; Lt. Col. Curtis A. Nelson to Maj. R.J. Coakley, Aug.
8, 1946, both in File MAN 000.7, Box 104903, CFU, FRCEP.
22. "Atom Plant Poll Inconclusive," *Business Week*
(Aug. 31, 1946), 77; "AFL Loses Prize," *Business Week* (Sept.
21, 1946), 101.
23. Maj. Harry C. Flowers to Mrs. J.W. Pleasant, Oct. 30,
1946, File MAN 080, Box 141909, CFU, FRCEP.
24. Col. J.S. Hodgson to File, Jan. 3, 1946, File MAN 334,
Box 104884 (33); Lt. Col. D.G. Williams to Maj. W.C.
Young, Aug. 30, 1946, File Oak Ridge Organization Study, Box
158535, both in CFU, FRCEP.
25. Clinton Hernandez to Employees of CEW Bus Authority, Jan. 23, 1945, File CEW 532, Box 381640, CFU, FRCEP.
26. Typescript of speech by unidentified Corps officer to
Oak Ridge Kiwanis Club, Sept. 7, 1946, File MAN 080, Box
141909, CFU, FRCEP.
27. Ibid.
28. Maj. John Turner to R.E. Stevens, May 1, 1946, File
680.42, Box 104885; L.Z. Dolan to Fred W. Ford, March 29,
1949, File 18, Box 158577, both in CFU, FRCEP.
29. First Lt. C.T. Vettel to Lt. Col. D.G. Williams, File
010729.3, Box 158535, CFU, FRCEP.
30. Lt. Col. John Hodgson to George Warlick, Aug. 27,
1946, File MAN 080, Box 104883, CFU, FRCEP.

31. Typescript of speech by unidentified Corps officer to Oak Ridge Rotary Club, 1946 [fall], File MAN 080, Box 141909, CFU, FRCEP.

32. "Civilian Commission to Visit Oak Ridge In Very Near Future," *Oak Ridge Journal*, Nov. 7, 1946; "Atomic Energy Commission Here," *Oak Ridge Journal*, Nov. 14, 1946.

33. *Oak Ridge Journal*, Jan. 2, 1947.

34. "Atlanta Hotel Fire Toll—114," *Knoxville News-Sentinel*, Dec. 7, 1946; "Atomic Commission Takes Ridge Reins: General Groves Quits," *Knoxville News-Sentinel*, Jan. 1, 1947.

35. *Knoxville Journal*, Jan. 1, 1947.

36. Maj. Gen. L.R. Groves to P.H. Hemphill, Dec. 3, 1946, File CEW 230.02, Box 156205, CFU, FRCEP.

EPILOGUE

1. Robinson, *The Oak Ridge Story*, 22, 25.

2. Interview #5, March 24, 1976.

3. Ford and Peitzsch, 96-97.

4. Ibid.; "Flattop Era at an End," *Oak Ridger*, July 15, 1957.

5. Interview #10, March 25, 1976.

6. "The Girls of Early Beacon Hall," *Oak Ridger*, March 22, 1973.

7. Interview #8, March 24, 1976.

8. Ford and Peitzsch, 51.

9. Interview #20, July 29, 1976.

10. *Report to the Atomic Energy Commission on the Master Plan, Oak Ridge Tennessee* (Skidmore, Owings & Merrill, 1948), 35.

11. *Oak Ridger*, Feb. 16, 1977.

12. *Oak Ridger*, Feb. 17, 1977.

13. Interview #30, June 26, 1976.

Bibliographical Essay

Because the three communities of Hanford, Washington; Los Alamos, New Mexico; and Oak Ridge, Tennessee, created by the Manhattan Engineer District were unique entities, very little secondary literature proved of significant value in research for this volume. Two studies of the domestic scene during World War II were especially useful in placing Oak Ridge in the wider context of American society at war: Richard R. Lingeman, *Don't You Know There's a War On?* (New York, 1971), and Richard Polenberg, *War and Society: The United States, 1941–45* (New York, 1972). General reading in historical works dealing with urban development and community dynamics provided the authors with valuable theoretical insights as well as a broader perspective on the Oak Ridge experience than might otherwise have been the case. Representative, and the most noteworthy, of these studies were the following: Maurice Stein, *Eclipse of Community* (Princeton, 1972); Rosabeth Kanter, *Commitment and Community* (Cambridge, 1972); Robert Dykstra, *The Cattle Towns* (New York, 1970); Stanley Buder, *Pullman* (New York, 1967); Gilbert Osofsky, *Harlem: The Making of a Ghetto* (New York, 1966); Robert and Helen Lynd, *Middletown* (New York, 1929) as well as *Middletown in Transition* (New York, 1937); Herbert Gans, *The Levittowners* (New York, 1967); Joseph Arnold, *The New Deal in the Suburbs* (Columbus, 1971); and Paul Conkin, *Tomorrow a New World* (Ithaca, 1959).

Within the category of general secondary reading, three studies merit special mention, though here too they provided little in the way of models against which to place the MED history of Oak Ridge. One was Arthur Maass' general history of the Corps of Engineers operations, especially in the area of flood control, *Muddy Waters* (Cambridge, 1951). A second was Lowell Carr and James Sterner, *Willow Run: A Study of Industrializations and Cultural Inadequacy* (New York, 1952), which well described the destructive effects of poor quality, overcrowded housing in a rural area hit suddenly by wartime industrial expansion. While the federal government was deeply involved in the Willow Run bomber plant and construction of support housing, there was minimal federal involvement in community development and virtually no military presence in the area on a long-term basis. Robert J. Havighurst's and H. Gerthon Morgan's *The Social History of a War-Boom Community* (New York, 1951) provided a valuable examination of the sudden impact of thousands of war workers on a small town, in this case Seneca, Illinois.

A small number of secondary volumes—published memoirs, and certain other miscellaneous unpublished works—did offer direct coverage of the Manhattan project. Some also yielded pertinent information, in varying amount as well as quality on the Oak Ridge experience. One such work was Richard G. Hewlett's and Oscar E. Anderson's thorough and perceptive overview of the entire MED project, *A History of the Atomic Energy Commission: The New World, 1939–1946* (University Park, Penn. 1962). The chapter on the Manhattan experience in Lenore Fine and Jesse Remington, *The Corps of Engineers: Construction in the United States* (Washington, D.C., 1972), proved a useful if brief reference and continued the high level of scholarship that characterizes other volumes in the United States Army in World War II series. Of lesser value was Anthony Brown and Charles MacDonald, eds., *The Secret History of the Atomic Bomb* (New York, 1977). It contained limited information on all three MED communities, though was largely concerned with the technical side of the atomic project. Stephane Groueff's *Manhattan Project: The Untold Story of the Making of the Atomic Bomb* (Boston, 1967) was also concerned primarily with the technical aspects of bomb development but presented a fascinating look at problems faced by scientists in its development as well as the individuals who worked to solve them. In their *Enola Gay* (New York, 1978) Gordon Thomas and Max Witts spoke not at all to the three MED communities but was highly dramatic and brilliant reconstruction on almost a day-by-day account of events, both American and Japanese, leading up to the Hiroshima bombing.

Of more direct significance was Volume Twelve of the unpublished thirty-five volume documentary chronicle entitled, "Manhattan District History." Prepared by the Corps of Engineers at the end of the MED period as an official record, these volumes offered a wealth of statistical data, though little in the way of interpretive narrative on the entire atomic project. Volume Twelve was devoted entirely to the Clinton Engineer Works and was located in Box 381638, CFU, FRCEP. In Chapter 1, Fred Ford and Fred Peitzsch, "A City is Born," dealt directly with the MED experience in Oak Ridge, and it was most informative. This work was written by two employees of what was then the Atomic Energy Commission at the request of that agency. Eighteen copies were produced and these were only released in 1978. Each copy included a stamped disclaimer indicating that the work did not carry official sponsorship by the Department of Energy.

Graduate students at the University of Tennessee in Knoxville explored in the 1950s two aspects of Oak

Ridge history and its relationships to surrounding areas. One was William A. Kelly, "The State and County Governmental Relationships of Oak Ridge, Tennessee" (M.A. thesis, 1951). The second was Dale E. Case, "Oak Ridge: A Geographic Study" (Ph.D. dissertation, 1955). Both provided useful insights into the MED years; the latter much more so than the former. Professor June Adamson of the University of Tennessee, Knoxville, College of Communications, reviewed the early years of the Oak Ridge newspaper in "From Bulletin to Broadside: A History of By-Authority Journalism in Oak Ridge, Tennessee," *Tennessee Historical Quarterly*, XXXVIII, No. 4 (Winter 1979), 479–93.

Halfway between secondary study and memoir was George Robinson's chatty *The Oak Ridge Story* (Kingsport, Tenn., 1950). Far from scholarly in approach, and perhaps most valuable for its anecdotal material, the volume provided a highly informal and patriotic account of reservation life from the perspective of the author's experiences as an Army officer stationed in Oak Ridge during the war period. Stone and Webster Engineering Corporation in *A Report to the People* (Boston, 1946) offered valuable information on the massive wartime construction effort by that company at Oak Ridge. Representative memoirs of individuals, valuable in varying degree, who were connected with the Manhattan project were: Leslie R. Groves, *Now It Can Be Told: The Story of the Manhattan Project* (New York, 1962); Arthur H. Compton, *Atomic Quest: A Personal Narrative* (New York, 1966); and Laura Fermi, *Atoms in the Family: My Life With Enrico Fermi* (Chicago, 1954), all of which yield only fragmentary glimpses of Oak Ridge.

Not surprisingly, virtually nothing on the community appeared in popular periodicals until August 1945 when there occurred a small flood of articles. Beyond fairly regular but brief news coverage in the immediate postwar period in *Time*, *Newsweek*, and *U. S. News and World Report*, a number of pieces were sufficiently useful so as to deserve special mention. A thoughtful thumbnail biography of Leslie Groves and his work was Robert DeVore, "The Man Who Made Manhattan," *Collier's*, October 13, 1945. Less well-known scientists were presented in Louis Falstein, "Men Who Made the A-Bomb," *New Republic*, November 26, 1945. *New Republic* also carried a second perceptive article by Falstein on Oak Ridge itself titled "Oak Ridge: Secret City," November 12, 1945, and much later a chilling piece by a former editor of the *Oak Ridge Journal*, Richard B. Gehman relative to the congressional investigations of some of those scientists in Oak Ridge who had fallen afoul of Washington's Cold War fears: "Oak Ridge Witch-hunt," *New Republic*, July 5, 1948. LeRoy A. Sheetz's article in *The Christian Science Monitor*, "Richland-the Atomic City," January 18, 1947, gave one of the few comparable writeups on "the other" atomic city.

Two especially thoughtful articles in the *New York Times Magazine* were: "A-Bomb's Home," by Warner Ogden, April 14, 1946; and "Oak Ridge Ponders the Future," by Cabell Philips, April 13, 1947. *Business Week* had extensive coverage of the organizing struggle between the AFL and the CIO from June through August 1946 and January 18, 1947, but most useful for this study was a lengthy article based on careful interviewing, "The City Atoms Built," October 27, 1945. Daniel Lang, who lived in Oak Ridge during the war, described his life in the city in two articles for the *New Yorker*: "Atomic City," September 29, 1945, and "Career at Y-12," February 2, 1946. An insightful, if somewhat patronizing view of Oak Ridge by one of the many Yankees who came south, was Theodore Rockwell, "Frontier Life among the Atom Splitters," *Saturday Evening Post*, December 1, 1945. Except for a two-page article on August 20, 1945, the only article in *Life* that dealt to any extent with the city was "Oak Ridge," by Francis Sill Wickware, September 9, 1946.

A number of specialized periodicals published, in

the immediate postwar period, articles relative to Oak Ridge that spoke from their own particular interest areas. These included: Elizabeth Edwards (supervisor of libraries for the Oak Ridge Recreation and Welfare Association), "Books on the Bomb Site," *Wilson Library Bulletin*, October 1945; Joan Lowrie, "Library-Nerve Center of the Schools," *Wilson Library Bulletin*, February 1946; Elizabeth Edwards, "City of the A-Bomb Has Library," *Library Journal*, September 15, 1945; Lester N. Recktenwald, "Guidance and Personnel Services in a 'War-Industry' Community," *School and Society*, January 26, 1946; "Taylor Instrument," *Fortune*, August 1946; "Atom City," *Architectural Forum*, October 1945; "How the Oak Ridge Street Program Grew," *American City*, April 1948; J. D. Robinson and T. R. Jarrell, "Recreation in America's Secret City," *Recreation*, February 1946; "Peace Among the Churches," *Christian Century*, August 29, 1945, and "Peace Among the Churches: A Reply," by Robert F. Lundy, *Christian Century*, September 19, 1945.

Two major collections of primary materials provided the major body of sources on which this volume is based. One was on deposit at the United States Federal Records Center in East Point, Georgia. This material, which was referenced under Record Group 326, Central Files Unclassified, contained approximately two thousand boxes of documents accumulated by the Manhattan Engineer District and its successor, the Atomic Energy Commission, in Oak Ridge. The second collection contained about two hundred boxes of community materials and remains currently in the custody of the Department of Energy at Oak Ridge. This collection was referenced under the heading, Central Files Classified and, while now largely declassified, bore various security classifications at the time the present study was begun. With the exception of Oak Ridge police records which were destroyed by the AEC, all these records are now readily available at the Oak Ridge offices of the Department of Energy.

Of much less value, though not without some insights into the Oak Ridge experience were at least two manuscript collections located in the National Archives at Washington, D.C. One was the Leslie Groves papers in Record Group 200. These papers were largely concerned with MED operations at the highest levels of administration and dealt only marginally with activity at Oak Ridge. More pertinent to developments in Tennessee, though still of limited value, were Records of the Office of the Chief of Engineers of MED in Record Group 77. This collection contained an extensive number of photographs of the Oak Ridge reservation during the first three years of its development.

Because of the tight security which characterized Oak Ridge in the war period, newspapers in adjacent areas provided limited insights into happenings on the federal reservation. Nonetheless, a careful examination of both the *Knoxville News-Sentinel* and *Knoxville Journal*, as well as the *Clinton* (Tennessee) *Courier-News* was necessary for the present study and did yield occasional comments about CEW from the perspective of these towns. A full run of all three journals is available on microfilm at the main library of the University of Tennessee, Knoxville. The reservation's weekly newspaper, the *Oak Ridge Journal*, also available on microfilm at the University, was more useful. For obvious reasons, national newspapers such as the *New York Times* were of very little value until after the war.

Finally, the authors conducted one-hour interviews with approximately seventy-five persons who had lived in Oak Ridge during and immediately after the war. Most of these individuals were members of a local organization known as "The '43 Club" whose members all came to the reservation in 1943. While the authors sought to obtain in these interviews a social and economic cross-section of residents, numerically the sample does in fact have a bias toward white-collar professionals. Where interview material was used in the text of this volume, it was footnoted by number and date of interview rather than by the names of those with whom we talked. This has been done to protect the privacy of our interviewees. The tapes have been transcribed and would be available through the authors to those whose scholarly research might be enhanced by examination of them.

Index

Acme Credit Corporation, 150. *See
also* Security
Alamogordo (New Mexico), 158. *See
also* Los Alamos; "Trinity Tests"
Alexander Motor Inn, 197
Aluminum Company of America
(ALCOA), 56
American Federation of Labor
(AFL): competes with CIO,
182–84; efforts at CEW, 177, 179,
180–84; mentioned, 180, 182. *See
also* Unionization efforts
American Industrial Transit, 186
Anderson County (Tennessee):
attitude toward surrounding area,
50–52, 206–7, 209; decision to
locate CEW, 6, 7, 41–49; impact of
CEW, 47, 52, 60–64, ch. 2; local
suspicion of CEW, 64; political
rivalry with Oak Ridge, 63, 176
Andrew Johnson Hotel, 192
Arkansas Cafeteria, 36. *See also* Oak
Ridge
Association of Oak Ridge Engineers
and Scientists, 177
Atomic Energy Act (1946), 168
Atomic Energy Commission (AEC):
and Oak Ridge blacks, 210–12;
assumes jurisdiction over CEW,
189–91; mentioned, 186, 189, 195
Atomic Trades and Labor Council,
180. *See also* Unionization efforts
Ayres, Eben, 162

Bacon Mills, 55
Baker, A.L., 105

"Battle of the Bridges," 61
Bayonne Hall, 200
Beacon Hall, 200, 201, 202
Beaumont Hall, 200
Bethel Valley Road, 215
Big Ed's Pizza, 202, 204
Black Oak Ridge, 10, 195
Blacks, Oak Ridge: "Negro Village,"
22, 111–12, 173, 210–14;
mentioned, 22–23, 104–5, 111–15,
117–18, 173–75; proposed family
housing, 173. *See also* Scarboro
Blair Gate, 10
Bloch, E.J., 68, 113
Blue Moon Cafe, 8
Bonnet, William, 169
Bonneville Dam (Washington), 6
Boone Hall, 201
"Brass Hat Circle," 105. *See also*
Oak Ridge
Bush, Vannevar, xix
Business Week, 171
Byrnes, James F., 156

Camp Forrest, 149
Canton Hall, 200
Carter, Jimmy, 205
"Castle, The," 197
Cemesto homes, 21, 100, 196
Central Facilities Advisory Council
(CFAC), 145
Chapel on the Hill, 129, 197, 198
Charleston Hall, 200
Chicago Defender, 175
Churchill, Winston, 159
Civilian War Housing Authority, 15

243

INDEX

Hutments (cont.)
212–14; mentioned, 12, 23, 87, 89,
100, 104–05, 111, 113, 117, 199
International Brotherhood of
Electrical Workers (IBEW), 177,
183. *See also* Unionization
efforts
International Brotherhood of
Fireman and Oilers (IBFO), 177.
See also Unionization efforts

J.A. Jones Construction Company,
186
Jackson Square, 36, 89, 198, 199,
200, 202. *See also* Oak Ridge
Jefferson Recreation Hall, 36
Jefferson Square, 32
Jennings, Congressman John, 57

K–25 (gaseous diffusion facility), 73,
75, 79, 106, 148, 156, 184, 193, 205
Kaiser Wilhelm Institute, xvii
Kellex Corporation, 8
Kelly, Wilbur, 12, 14
Kingsport (Tennessee), xxi
Kingston(Tennessee), 45, 60
Kingston Demolition Range, 8, 207.
See also Clinton Engineer Works
Knox County (Tennessee), 56, 61,
62, 75
Knoxville (Tennessee): comments
on CEW by *Journal*, 190,
News-Sentinel, 190, 193;
mentioned, 8, 31, 45, 50, 52, 57,
58, 59, 60, 63, 79, 121, 142, 164,
166; Oak Ridge prices based on,
24; unionization efforts at CEW
from, 180; reaction to Hiroshima
bombing, 164
Kokura (Japan), 159, 160

Leatherman and Alley Company,
186
"Lee-Ross Study," 115
Leonard, G.B., 49
Lilienthal, David, 190
Lincoln Cafeteria, 36
"Little Boy" bomb, 159, 160
Los Alamos (New Mexico), xix, xxi,
158, 183
Louisiana Cafeteria, 36

McKellar, Senator Kenneth D., 55

Manchester, William, 162
Manhattan Engineer District (MED):
bomb production initial
schedule, 155; Hiroshima
bombing, 160, 162, 163, 164;
housing regulations at Oak Ridge,
101–7; land acquisition in
Tennessee, 39, 41, 43, 45, 99;
legacy at Oak Ridge, 199, 204–6,
215, 216; lifestyle at Oak Ridge,
ch. 4; memory of, 193–94;
mentioned, xix, xxi, xxii, 4;
mission of, 30; final phase of,
159–62; view of, 99,
"normalization" efforts at, 35–38,
65, 184–89; Oak Ridge and
postwar administration, ch. 6;
pre-MED population removal,
41–45; resident life, involvement
in, 134–35; and Roane-Anderson
Company, 67–70, ch. 3; search for
initial location, 6, 7; and security,
ch. 5; terminated, 190; townsite,
original supervision of, 65, 66;
plans for, 14, 15, 16, 35; view of,
156, 177, 179, 180–85; and
unionization efforts at CEW, 183.
See also Oak Ridge
Marshall, James C., 12, 16, 17, 21,
36, 43
Maryville (Tennessee), 212, 213
Mayflower Restaurant, 200
Memphis Commercial Appeal, 158
Merrill, John, 14, 16
Middletown Center, 32. *See also*
Oak Ridge
Midtown Trailer Park, 22, 199
Miller's Department Store, 60
Monsanto Chemical Company, 10,
184

Nagasaki (Japan), xvii, xix, 159, 160,
162, 164
National Commission on Civil
Rights (1946), 175
National Defense Research
Council, xvii
National Labor Relations Board,
177
"Negro Village," 22, 111, 112, 173,
211. *See also* Blacks
New Deal, xxi, 49
Nichols, Kenneth D., 106, 167, 168